Y0-ARM-982

McGraw·Hill
Learning Materials

SPECTRUM
GEOGRAPHY
UNITED STATES OF AMERICA

Grade 5

Authors

James F. Marran
Social Studies Chair Emeritus
New Trier Township High School
Winnetka, Illinois

Cathy L. Salter
Geography Teacher
Educational Consultant
Hartsburg, Missouri

McGraw·Hill
Learning Materials

8787 Orion Place
Columbus, OH 43240-4027

EAN

The **McGraw·Hill** Companies

9 781577 681557

90000

Program Reviewers

Bonny Berryman
Eighth Grade Social Studies Teacher
Ramstad Middle School
Minot, North Dakota

Grace Foraker
Fourth Grade Teacher
B. B. Owen Elementary School
Lewisville Independent School District
Lewisville, Texas

Wendy M. Fries
Teacher/Visual and Performing Arts Specialist
Kings River Union School District
Tulare County, California

Maureen Maroney
Teacher
Horace Greeley I. S. 10 Queens
District 30
New York City, New York

Geraldeen Rude
Elementary Social Studies Teacher
1993 North Dakota Teacher of the Year
Minot Public Schools
Minot, North Dakota

McGraw-Hill
Consumer Products
A Division of The McGraw-Hill Companies

Copyright © 1998 McGraw-Hill Consumer Products.
Published by McGraw-Hill Learning Materials, an imprint of McGraw-Hill Consumer Products.

Send all inquiries to:
McGraw-Hill Learning Materials
8787 Orion Place
Columbus, OH 43240-4027

ISBN 1-57768-155-X

Table of Contents

Introduction to Geography

The Land Prior to Columbus

The Age of Exploration

The Colonial Period

Lesson 7

Lesson 8

The Revolution and Founding of a Nation

Lesson 9

Lesson 10

Westward Expansion

Lesson 11

Lesson 12

1

Lesson 1

Where on Earth?

As you read about the continents and oceans that make up Earth, think about where you live on Earth.

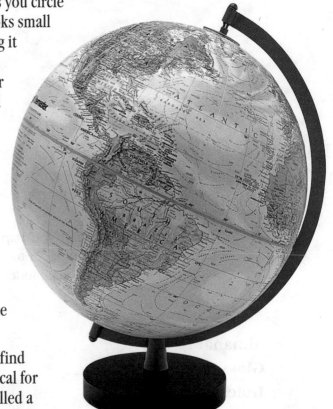

Space travelers are treated to this spectacular view of our planet.

Imagine you are an astronaut looking down on Earth as you circle the planet in your spacecraft. From this height, Earth looks small and round. White clouds swirl around the planet, making it resemble a marble. The deep shades of blue that you see are mainly water. Earth has four main bodies of water, or oceans. They are the Pacific, the Atlantic, the Indian, and the Arctic oceans. The oceans are connected to form one vast body of water that covers much of Earth.

In places you can see large land masses interrupting the oceans. These land masses are the seven **continents** of Earth. The continents are North America, South America, Africa, Asia, Europe, Australia, and Antarctica. Most of Earth's people make their homes on the continents. The people of the United States, for example, live on the continent of North America. Nigerians live in Nigeria, a country on the continent of Africa.

Viewing Earth from an orbiting spacecraft is one way to find out what our planet looks like. However, that isn't practical for most of us! A better way is to look at a model of Earth called a **globe.** It shows the correct shape of Earth's major land masses and bodies of water.

A globe is shaped like Earth.

A globe, like the Earth it represents, is a **sphere**–that is, it's shaped like an orange. Holding and turning a globe gives you a number of views of Earth. With a globe, you can see at one time what half of Earth looks like.

Although a globe is the best way to represent all of Earth at once, globes are not handy to carry around. Also, Earth is so large that a globe can show few details of its surface. For these reasons, geographers make flat drawings of Earth, and of parts of Earth, called **maps.** On a flat world map, you can see all of Earth at once. And flat maps are easy to carry and store!

When mapmakers create flat world maps, they often change Earth's size and shape. Earth's oceans and continents appear out of shape. Some are the wrong size in relation to others. You can understand why by noticing what happens when you cut an orange peel and try to flatten it, as in the illustration.

World maps, globes, and photos taken from outer space are all valuable tools. They provide us with different ways of looking at the Earth on which we live.

This world map is another way of looking at Earth.

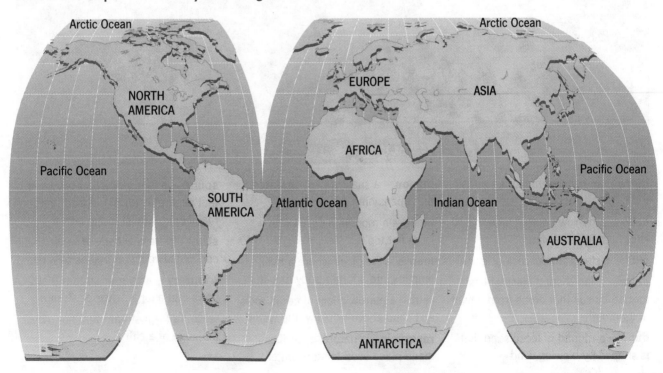

Arctic Ocean

NORTH AMERICA

Pacific Ocean

SOUTH AMERICA

Atlantic Ocean

EUROPE

ASIA

AFRICA

Indian Ocean

Pacific Ocean

AUSTRALIA

Arctic Ocean

ANTARCTICA

Lesson 1
MAP SKILLS Identifying Landforms and Water Forms

Specific terms are used to identify Earth's landforms and water forms.
The below diagram shows some of these terms.

butte – an isolated mountain or hill with a small flat top

cape – a narrow point of land that sticks out into the water

cliff – a steep face of rock or earth

delta – a fan-shaped piece of land formed by soil that drops from a river at its mouth

dune – a mound of loose sand that is shaped by blowing wind

lagoon – a shallow body of water that opens onto the sea

mesa – a mountain or hill with a large flat top

mouth – the place where a river empties into a larger body of water

reef – a narrow chain of coral, rock, and sand above or below the water

reservoir – a natural or artificially made place used to store water

sound – a long, wide body of water that separates an island from the mainland; parallel to the coast

strait – a narrow body of water that connects two larger bodies of water

tributary – a river or stream that flows into a larger river or stream

waterfall – a stream that flows over the edge of a cliff

A. Using the diagram, list the different forms of land and water in the appropriate columns below.

Landforms	Water Forms
butte	lagoon
cape	mouth
cliff	reef
delta	volcano
lava	sound
mesa	waterfall
?	strait

B. Identify the landforms and water forms described below and complete the puzzle.

1. A stream that flows over the edge of a cliff. W_ _ _ _ _ _ _ _ (waterfall)

2. A narrow body of water that connects two larger bodies of water. _ _ _ _ _ _ _ (strait)

3. A river or stream that flows into a larger river or stream. _ _ _ _ _ _ _ _ _ (tributary)

4. The place where a river empties into a larger body of water. _ _ _ _ _ (mouth)

5. A point of land that extends into the water. _ _ _ _ (cape)

C. The following landforms and water forms are also on the map, but are not labeled. Label them.

1. mountain   6. volcano
2. river   7. glacier
3. ocean   8. bay
4. plateau   9. canal
5. hill   10. valley

D. Turn to the world map in the Almanac. How many other forms of land and water can you find?

Lesson 1

ACTIVITY

Identify where you live on Earth and the forms of land and water in your community.

The Right Spot

The **BIG** Geographic Question

Where do you live and what landforms and water forms are found there?

In the article you were reminded of Earth's seven continents and four oceans. In the map skills lesson you learned to identify some of the landforms and water forms on Earth. Find out which of these landforms and water forms are in the area where you live.

A. Tell what you know about where you live.

1. In what continent and country do you live?

   North America, United States

2. What is the name of your state?

   New York

3. What is the name of your community?

   Flushing

B. Think about what your state looks like.

1. Does your state have many lakes? _No_

2. Is it near one of the four oceans? _Yes_

3. Is your city nestled in a valley or located high in the mountains? _No_

4. Write what you know about your state's landforms and water forms.

   _____

   _____

   _____

C. Now think about your community and answer the following questions.

1. Is your community a city, a town, a village, or a suburb?

   _____

2. How many people do you think live in your community?

   _____

3. Put a check beside each of the landforms and water forms listed below that are located in or near your community. Add to the list if necessary.

_____✓_____ lake	_____ waterfall
_____ river	_____✓_____ island
_____ mountain	_____ hill
_____ plain	_____ valley
_____ plateau	_____ cape
_____ canal	_____✓_____ bay
_____✓_____ ocean	_____ cliff
_____ dam	_____ coast
_____ harbor	_____ desert

4. Draw pictures or use construction paper or magazines to cut out pictures of landforms or water forms found in or near your community.

D. Make a mobile to show some of the physical features that are found where you live on Earth. Attach the drawn or cut-out pictures that you have collected to pieces of string cut at different lengths. Tie the string to a hanger to make a mobile. Compare your mobile to your classmates' mobiles.

Lesson 2

Getting a Map's Message

As you read about the features of a map, think about how each feature helps communicate a map's message.

Have you ever had trouble figuring out a map? A map's main features give you the clues you need to figure it out!

Your first clue is the map's **title.** The title describes the main image, the place or region, shown on the map. It also sums up the map's main idea. Sometimes the title might include a date to show the time period of the information on the map.

Other clues are sprinkled all over the map. The words or phrases that identify the features on a map are called **labels.** They tell you the names of rivers, towns, mountains, buildings, and other features.

Another important part of a map is the **compass,** which reminds you which direction is north, south, east, and west and helps you find your way around a map. Although not all maps include a compass, a person reading a map is usually aware of the directions in which various features lie on a map.

A **map scale** is another clue that can help you figure out information on a map. It shows how many miles or kilometers the inches or centimeters on a map represent.

A PLAN
of the several Villages in the
ILLINOIS COUNTRY,
with Part of the
River Mississippi &c.
by
Tho.ˢ Hutchins.

Scale of Miles.
0 1 2 3 4 5   10   15   20

This is a copy of a 1778 map. It shows planned villages during that period in history.

8

A map might seem even more mysterious when you learn that it has a legend. The **legend**, or key, lists the map's symbols and what they stand for. It usually has a box around it. Just remember that you can unlock a map's mysteries with the key in the box!

The title, labels, compass, map scale, and legend are important tools to help you unlock the message of a map. When approaching any map, it is a good idea to first look at these important parts individually, then put together all of the information that they provide to get the full message of the map.

The look of a compass can vary a great deal. The compass on the left is from a map made in 1676. The compass on the right is a more simplified one.

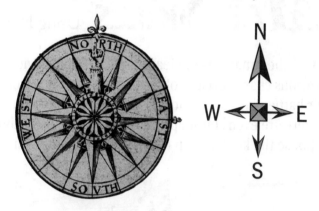

This map's legend shows the jobs of people living in New York City in 1703. The mapmaker has used clever symbols for different workers.

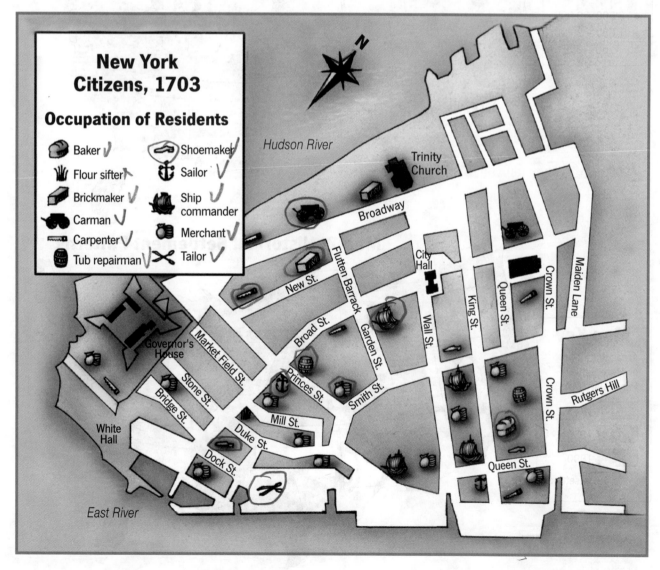

Lesson 2
MAP SKILLS Using Maps to Show Information

With their many useful features, maps can provide a lot of information about various characteristics of Earth. They can show routes for car trips, combat zones or battle plans, or the location and elevation of the world's highest mountains. There are even star maps that chart happenings above Earth. Mapmakers have to choose the kind of map that is best for the information they want to communicate.

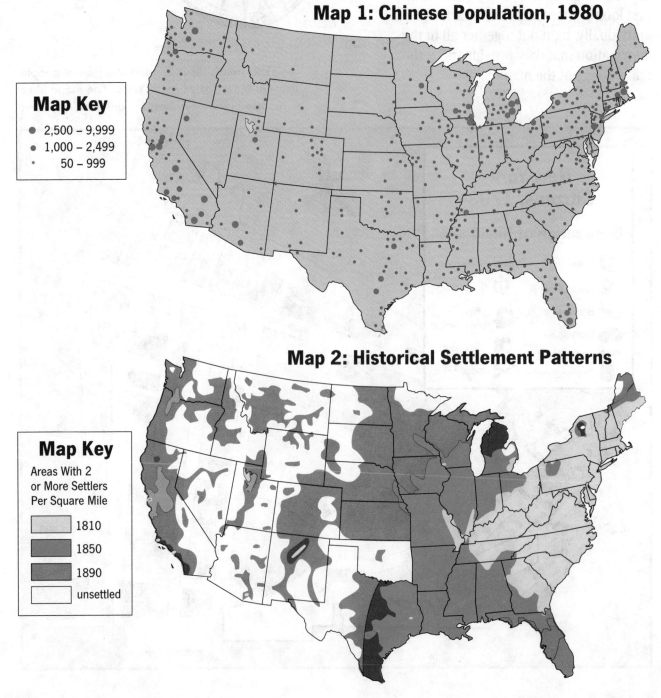

Map 1: Chinese Population, 1980

Map Key
- ● 2,500 – 9,999
- ● 1,000 – 2,499
- · 50 – 999

Map 2: Historical Settlement Patterns

Map Key
Areas With 2 or More Settlers Per Square Mile
- 1810
- 1850
- 1890
- unsettled

A. Look at the maps and then complete the chart below by writing a brief description of the information you see on each map.

Map#	Description
1	
2	

B. Read the following definitions of some of the different kinds of maps. Then write the name of the correct type of map next to the map numbers below.

Types of Maps and Their Definitions

- Political map—a map that shows borders of states and countries and the location of capitals and other cities or towns
- Physical map—a map that shows physical features, such as mountains, rivers, oceans, and deserts
- Historical map—a physical or political map that shows borders, physical features, or events of the past
- Population map—a map that shows how many people live in a place
- Road map—a map that shows the streets, roads, or highways that can be used to get from one place to another

Map#	Type of Map
1	Chinese population, 1480
2	Historical settlement patterns

C. Look at the two other types of maps in the Almanac on page 115. What types of maps do you think they are? Describe or define each one.

Map#	Type of Map	Definition
1		
2		

D. Look at the maps on pages 8 and 9.

1. What type of map do you think the Illinois Country map on page 8 is?

2. What type of map do you think the New York Citizens map on page 9 is?

Lesson 2

ACTIVITY
Make a map to communicate important information.

Make Your Own Map

The **BIG**
Geographic Question

What are the main features needed to create an effective map?

From the article you learned about the main features of a map. The map skills lesson helped you identify several different kinds of maps. Now make a map to communicate information to a friend or family member.

A. Under each of these headings, write titles for a map you might be interested in creating. Then put an **X** in the box next to the map you will create.

☐ **HISTORY** (family, state, city)
Example: How My Family Came to the United States

_____

☐ **EVENTS** (vacations, sports events, concerts)

_____

☐ **PLACES** (school, home, neighborhood, city, country)

_____

B. Answer the following questions about the map you are interested in creating.

1. What type of map will best communicate information about your topic? _____

2. What title best summarizes the map? _____

3. What main image will the map show? _____

4. What labels will be included on the map? _____

_____

5. What information will need to be included in the map's legend? _____

6. Does your map need a scale? _____

C. Now that you have come up with a written list of things to include on your map, decide how it will look.

   1. In the space below, sketch the symbols you will include in the legend on your map.

   2. Decide what the main image on your map will look like. Use the space below to work out a design for the map. Don't forget to include a compass if your map needs one.

D. Now that you have organized the parts of your map and made a sketch, create your final map on a separate piece of paper. When you have completed your map, show it to a family member, classmate, or friend. To see whether your map communicates what you wanted it to, have the person you share it with describe what they think the map is about.

Lesson 3
Ever-Changing Earth

As you read about the natural forces that have changed the shape and surface of North America, think about their influence on Earth as a setting for human life.

Today North America looks very different from the way it looked long ago before people settled the land. Natural forces both inside and outside of Earth keep changing the shape of the land. Over hundreds of millions of years, these forces have built up and worn down Earth's surface, creating landforms such as plains, plateaus, hills, and mountains.

Long periods of cold, known as Ice Ages, played a major role in changing the landscape of North America. Huge sheets of ice, called **glaciers,** moved southward from the Arctic like bulldozers, scraping up dirt and rocks from one place and pushing them to another. Much of the fertile soil in the midwestern United States is made up of layers of this scraped up dirt and rocks, or **till,** left behind by these moving, ancient ice sheets. These mighty glaciers formed the Great Lakes near Michigan, the Finger Lakes of New York, and the valleys of Yosemite National Park in California.

crust

mantle

outer core

inner core

Another process that is believed to have changed North America's landscape is **plate tectonics.** This theory, or explanation of what happened, grew from an earlier one about drifting continents. The plate tectonics theory suggests that Earth's outer crust is made of huge slabs of rock called plates. The upper mantle, which is liquid, moves the plates below Earth's surface in ways called "folding" and "faulting." This makes Earth's crust rise up and build mountains. Folding is a physical process that bends rock into a series of folds much like the pleats in an accordian. Faulting takes place when masses of rock in Earth's crust pull apart or push together. One plate of rock rides up over or slides under the other.

Mountains are the tallest landforms. They rise at least 2,000 feet above sea level. Mountains can also be formed by active volcanoes on land and on the ocean floor. Hills, landforms that are not as high or peaked as mountains, are lower than 2,000 feet above sea level.

Plateaus are the flat highlands, or pushed-up sections of Earth's crust, that sit like tabletops above the surrounding area. Plains, on the other hand, are large areas of lowlands–flat, rolling land. Many of the plains were formed by shallow seas that covered the land centuries ago.

The two forces that affect all of these landforms are weathering and erosion, caused by water, wind, and glacial ice. **Weathering** is the process of wearing down or change that occurs from being exposed to harsh climate. **Erosion** is the process of stripping away rock and soil from the surface of Earth and moving them to another place. Mountain rivers like the Colorado River carved deep canyons in the Colorado Plateau. Powerful ocean waves pounded the coasts of North America, forming inlets and tearing away beaches.

With all these natural forces at work, it is easy to understand why the geography of North America has changed greatly over the last two million years and is still changing today.

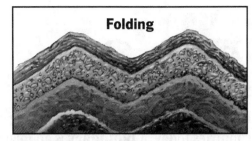

Folding

Faulting

Faulting

This diagram shows the basic landforms and water forms that have been carved out by Earth's forces.

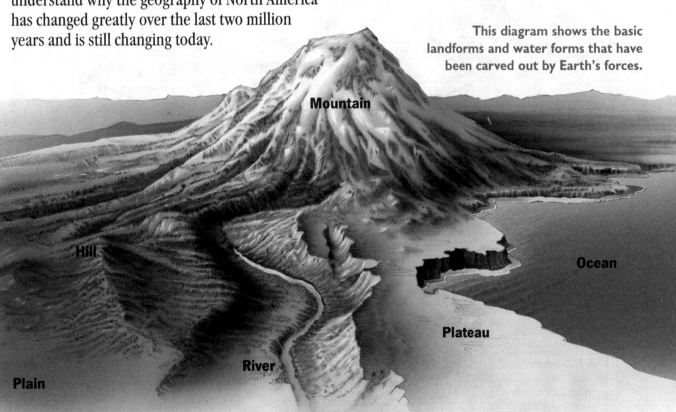

Mountain

Hill

Ocean

Plateau

River

Plain

15

MAP SKILLS Using a Map to Look at Boundaries

Forces that shaped North America long ago created many of the boundaries between
countries today. A **boundary** is an imaginary line that divides two states or countries.
Physical features, such as mountains, that divide land are **natural boundaries.**
Boundaries not formed by a physical feature are **artificial boundaries.**

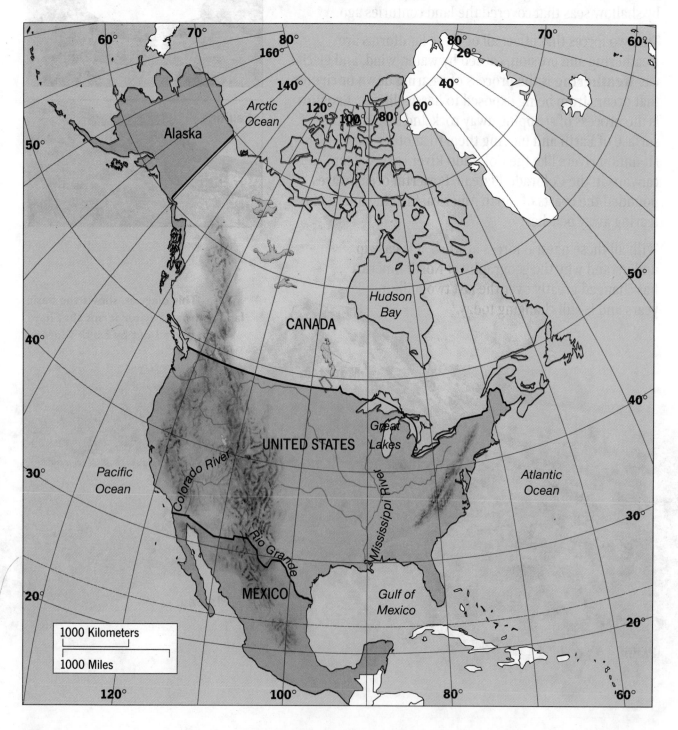

A. Look at the map of North America and answer the following questions.

1. What are the three countries of North America?

   Alaska, Canada and Mexico

   Canada

   Mexico

2. What name is used to refer to the five lakes that form a boundary between Canada and the United States? _Great Lakes Mississippi River ___ ___

   Is it a natural or artificial boundary? _____

3. Name the river that forms a boundary between Mexico and the United States. _Mississippi River _____

   Is it a natural or artificial boundary? _natural _____

4. Name the area of land that is on the northwestern boundary of Canada and the country that owns that land. _____

   Is this a natural or artificial boundary? _____

5. What is the number of the parallel line that forms a long section of the boundary between Canada and the United States? _____

   Is this a natural or artificial boundary? _____

B. Look at the coastline boundaries of North America and complete the following questions.

1. What ocean borders the east side of the United States and Canada? _Pacific_

2. What ocean borders the west side of North America? _Atla_

3. What ocean is north of Canada? _Artic_

4. What large body of water lies between Mexico and the Atlantic Ocean? _River_

C. Draw some conclusions about the above information.

1. What does the map of North America tell you about boundaries? _____

   _____

2. Why do you think waterways are common boundaries in North America? _____

   _____

Lesson 3

ACTIVITY Find out what North America looked like before it was first settled by people.

North America Long Ago

The **BIG** Geographic Question

How has the landscape of North America changed as a result of the last Ice Age?

From the article you learned how forces of nature shaped Earth. The map skills lesson showed you some of the natural boundaries of North America that these forces formed. Now find out how some of the physical characteristics that resulted from the last Ice Age have created big changes in North America's vegetation.

A. Take notes about what North America was like after the last Ice Age. Use the physical map of North America in the Almanac to find this information.

1. Land Notes—Many of the landforms we see today were created as ice moved across North America. Glaciers of long ago pushed and pulled at the land to shape the mountains and plains of today.

   a. Name the major mountain ranges of North America. _____

   _____

   b. Name the plains that are located along North America's eastern and

   southern coastlines. _____

   c. Name a major plateau in North America. _____

   d. What is the name of the large stretch of plain in central North America? _____

   _____

2. Water Notes—Several North American lakes and rivers were created as glaciers melted and retreated north as the Ice Age was coming to an end. Land and valleys carved by the moving ice filled with water as the ice melted.

   a. List the major lakes of North America. _____

   b. What is the longest river in North America? _____

c. What river runs along the boundary between the present-day countries of

Mexico and the United States? _____

B. Now that you have learned how the Ice Age changed the physical landscape of North America, use the vegetation maps in the Almanac to answer questions about vegetation patterns before and after the Ice Age.

1. **Vegetation Notes Before—Vegetation patterns were much different before the moving glaciers and melting ice changed the physical features of North America.**

   Where were the following vegetation regions located at the end of the last Ice Age?

   a. deciduous forests _____

   b. coniferous forests _____

   c. grasslands _____

   d. mixed forests _____

2. **Vegetation Notes Today—Patterns in the growth of vegetation in North America have changed as a result of the Ice Age. The presence of rivers has provided enough moisture for new types of vegetation to grow. Till, the fertile soil left behind as mountains and plains were formed, also allowed new types of vegetation to grow.**

   Where are the following vegetation regions located today?

   a. deciduous forests _____

   b. coniferous forests _____

   c. grasslands _____

   d. mixed forests _____

C. Based on all of the above information, explain the growth of present-day vegetation in North America. Think about how plants need fertile soil and moisture to grow. Include information on how the Colorado and Mississippi rivers have contributed to changes in vegetation.

Lesson 4

LIVING OFF THE LAND

As you read about the land of North America long ago, think about how Native Americans prior to the arrival of Columbus were affected by its geography.

The geography and climate of North America is incredibly varied. Before Columbus visited North America in 1492, Native Americans had already learned to use the resources of the land in the areas in which they lived.

For example, the eastern woodlands were dense with trees, and animal life thrived there. Forests stretched from north to south with rivers, lakes, and swamps between. Native American tribes such as the Iroquois hunted small and large game, fished, and gathered nuts and berries in the woodlands. They made clothing by weaving the bark of trees and preserving animal skins. They used trees to carve canoes from hollowed logs for transportation.

The Native Americans of the eastern woodlands built wooden dwellings called longhouses for shelter.

In contrast to the east, the southwest was mostly desert, with mountains and deep canyons. Native Americans could not rely on the few forests and animals available there. They raised crops—beans and corn for food and cotton to weave into clothing. Because the area had little rain, these Native Americans dug canals from rivers and lakes to water their crops. The sunbaked soil provided raw material for shelters.

Mixing mud with straw, tribes such as the Hopi built huge apartment-like complexes known as pueblos.

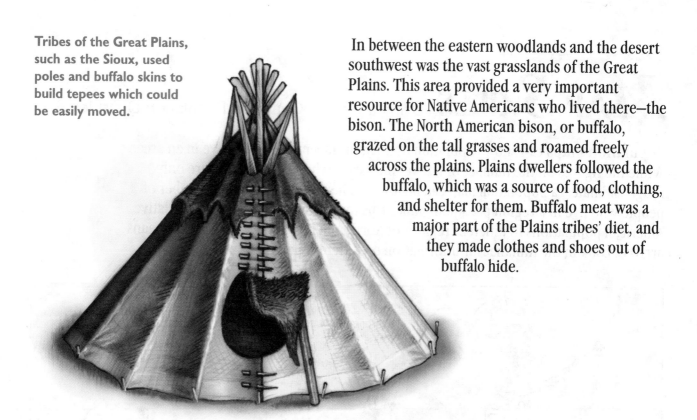

In between the eastern woodlands and the desert southwest was the vast grasslands of the Great Plains. This area provided a very important resource for Native Americans who lived there—the bison. The North American bison, or buffalo, grazed on the tall grasses and roamed freely across the plains. Plains dwellers followed the buffalo, which was a source of food, clothing, and shelter for them. Buffalo meat was a major part of the Plains tribes' diet, and they made clothes and shoes out of buffalo hide.

In the far west corner of North America, the peoples of the northwest did not rely on any one natural resource for their living. The wet climate sustained many resources for tribes such as the Chinook. The rivers and the Pacific Ocean were filled with fish. The mountains were thick with forests that provided timber for shelter and animals for food. Because trees were plentiful, people made almost everything of wood. They carved wooden canoes, which they used for fishing and for harpooning whales. These Native Americans of the northwest also wore clothing woven from cedar bark.

Native Americans of long ago learned to interact with their environments in order to survive. For the most part, they adapted their ways of living to the geography around them instead of trying to change their surroundings to supply their needs.

Native Americans of the northwest used the region's abundant timber to build plank houses, which they arranged in small villages along the coast.

21

Lesson 4
MAP SKILLS Using a Map to Compare Population Densities

Population maps can show population density, or how many people live in an area. While the figures on the number of Native Americans who lived long ago are only rough estimates, a population map can be used to show the approximate number of people who live in a region per square mile. This map compares the density of Native American populations in different regions of North America at the time the Europeans arrived. Look in the Almanac for information on different population terms.

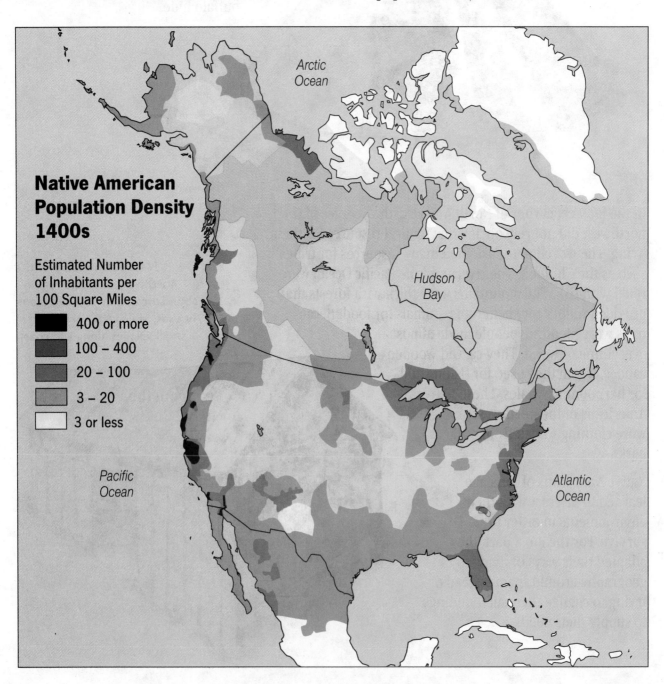

Native American Population Density 1400s

Estimated Number of Inhabitants per 100 Square Miles

- 400 or more
- 100 – 400
- 20 – 100
- 3 – 20
- 3 or less

A. Study the map of North America. Circle and label the following regions of the United States.

1. Northwest
2. Southwest
3. Eastern Woodlands
4. Great Plains

B. Compare the color patterns for high and low population densities on the map.

1. Which color represents a low population density? _____ yellow _____

2. Which color represents a high population density? _____ black _____

3. Which areas of the United States had a low population density? _____

_____

4. Which areas of the United States had a high population density? _____

_____

C. Use the map, the information on pages 20–21, and the Almanac to describe the physical features of the following areas.

1. Northwest _____

2. Southwest _____

3. Eastern Woodlands _____

4. Great Plains _____

D. Explain why some areas in the United States might have had an especially high population density. Base your answer on what you know about the geography and natural resources of the four regions prior to the time of Columbus.

_____

_____

_____

_____

_____

_____

Lesson 4

ACTIVITY Create alternate dwellings for Native American tribes.

Native American Dwellings

The **BIG** Geographic Question

What alternate dwellings could Native American tribes of long ago have built based on the geography and resources of the regions in which they lived?

From the article you learned how Native Americans of long ago adapted their ways of living to the physical geography and natural resources around them. The map skills lesson showed you where many or few Native Americans of long ago lived. Now describe and draw or create a model of an alternate dwelling for each of the following Native American tribes: the Makah, the Navajo, the Cheyenne, and the Seneca.

A. Use the Almanac to find the region in which each **Native American** tribe below lived.

1. Makah _____   **3.** Cheyenne _____

2. Navajo _____   **4.** Seneca _____

B. Use the article and Almanac to answer the following questions.

1. What were the physical features of the region in which each group lived?

_____

_____

2. What natural resources were available to each group?

_____

_____

3. What was the climate like in each region?

_____

_____

_____

24

C. Organize the information you have gathered onto the chart below. Use the article and Almanac to find any missing information.

Tribe	Region	Climate	Use of Resources	Types of Past Shelter	Drawing of Shelter
Makah					
Navajo					
Cheyenne					
Seneca					

D. Review the information on the above chart. Select one of the tribes and design an alternate dwelling the people of the tribe could have built using the resources they had. Explain how this structure would have been useful in the environment (climate and geography) of the region. Sketch your alternate dwelling on the chart. Make a model of the alternate dwelling.

Lesson 5

TOOLS FOR TRAVELING IN TIME

As you read about how people traveled in the fifteenth century, think about how technology affects where and how far people can go.

Europeans in the 1400s, or fifteenth century, had limited knowledge of world geography. The most influential maps of the time were those drawn by Ptolemy (TAHL-uh-mee), an ancient Greek astronomer and geographer. However, his maps showed mistaken information, such as Asia and Africa as a connected landmass and the Indian Ocean as an enclosed lake.

At first mariners, or sailors, rarely traveled very far from home. They sailed in known waters or on enclosed seas like the Mediterranean. Then the mariners began traveling south along the coast of Africa. They were unfamiliar with the coastline, and the people on the coast were not always friendly.

Navigators needed ways to determine their **latitude,** or distance from the equator. They invented instruments such as the **astrolabe,** the **quadrant**, and the **cross-staff** to help them. Navigators determined **longitude**, or distance east or west of their starting point, by estimating how far and in what direction they had traveled.

Maps were a valuable tool for exploration. They became more accurate thanks to the work of Prince Henry of Portugal, known as Henry the Navigator. He supervised the exploration of the west coast of Africa in the 1400s. Prince Henry sent out more than 50 expeditions and required sailors to bring back detailed notes and maps of their journeys. Portuguese mapmakers revised and improved the navigational maps after each voyage.

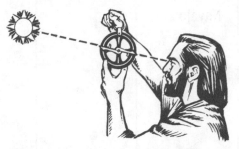

The astrolabe is used to measure the height above the sea of the celestial bodies such as planets and stars.

A quadrant is used in navigation and astronomy to measure distance above the horizon.

A cross-staff is used to determine latitude by measuring the height of the sun and stars above the horizon.

Explorers also needed ships to take them to a new part of the world and bring them home again. Merchants had been sailing in the Mediterranean in big sailing vessels that could hold hundreds of tons of cargo. In the mid-fifteenth century, Prince Henry's shipbuilders invented the caravel, a smaller ship that was better at sailing into the wind.

Today it may seem that fifteenth-century explorers had a long way to go in terms of transportation technology. But if you really think about it, they made many important discoveries that are still being used in the twentieth century.

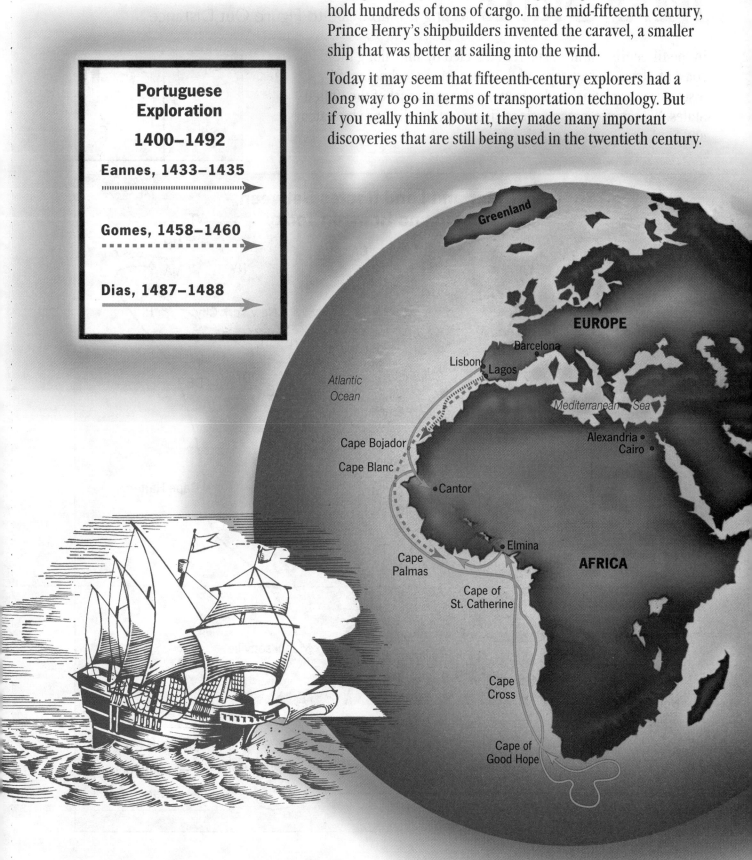

Portuguese Exploration

1400–1492

Eannes, 1433–1435

Gomes, 1458–1460

Dias, 1487–1488

Greenland

EUROPE

Atlantic Ocean

Barcelona

Lisbon • Lagos

Mediterranean Sea

Alexandria •
Cairo •

Cape Bojador

Cape Blanc

• Cantor

• Elmina

AFRICA

Cape Palmas

Cape of St. Catherine

Cape Cross

Cape of Good Hope

MAP SKILLS Using a Map Scale to Figure Out Distance

In the fifteenth century explorers traveled by ship along the coast of Africa. Today, some of the cities where technological discoveries are being made are along the east coast of the United States. We can use a map scale to figure out the distance by water and land, between some of the cities.

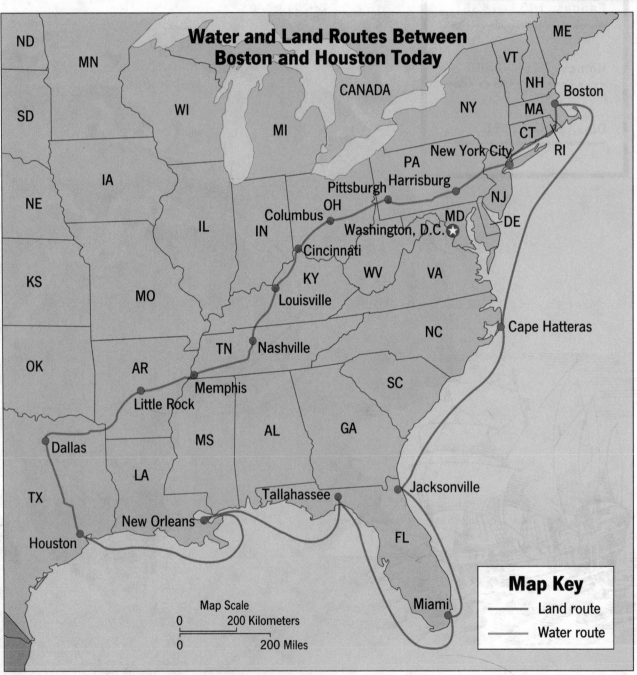

Water and Land Routes Between Boston and Houston Today

ND
MN
SD
WI
CANADA
ME
VT
NH
Boston
NY
MA
CT
RI
IA
NE
New York City
PA
Harrisburg
Pittsburgh
NJ
IL
Columbus
OH
MD
DE
IN
Washington, D.C.
KS
MO
Cincinnati
KY
WV
VA
Louisville
TN
Nashville
NC
Cape Hatteras
OK
AR
SC
Memphis
Little Rock
AL
GA
Dallas
MS
LA
Jacksonville
TX
Tallahassee
New Orleans
FL
Houston
Miami

Map Scale
0   200 Kilometers
0   200 Miles

Map Key
—— Land route
—— Water route

A. Use the map scale to answer the following questions.

1. How many miles are represented on the scale?

_____

2. Using a ruler and the map scale to determine the distance, approximately how many miles by air is Boston from New York City?

_____

B. Use a piece of string and a ruler to measure the distance by water between Boston and Cape Hatteras on the map and answer the following questions.

1. How many inches is the length of string that represents the distance from Boston

   to Cape Hatteras? _____ inch(es)

2. Measure the miles line on the map scale. Complete the following.

   _____ inch(es) is equal to _____ miles.

3. How many miles by water is Boston from Cape Hatteras? _____

C. Use a string and the map scale to measure and determine the distance of the land and water routes between Boston and Houston.

1. How long is the land route from Boston to Houston? _____

2. How long is the water route from Boston to Houston? _____

3. If you had been traveling in the 1400s from what is now Boston to what is now Houston, which route would you have taken? Think about the ways people traveled during that time. Explain your answer.

   _____

   _____

   _____

4. If you were traveling from Boston to Houston in the 1990s, would you take the water or land route? Explain why.

   _____

   _____

   _____

   _____

Lesson 5

ACTIVITY
Find out how travel time, distance, and technology are related.

How Far? How Long?

The **BIG** Geographic Question

How is the time it takes to cover a particular distance related to the transportation technology available?

In the article you learned what a difference technology made to explorers in the fifteenth century. The map skills lesson showed you how to use a map scale to determine the difference in land and water travel distance. Now find out how the means of transportation have continued to advance. How has technology helped people go farther in less time?

A. Gather information from the Almanac to complete the chart below.

Traveler and Date of Journey	From/To	Method of Travel	Distance	Time
Marco Polo (1271–1274)				
Christopher Columbus (1492)				
Ferdinand Magellan (1519–1521)				
Meriwether Lewis and William Clark (1804–1806)				
Charles Lindbergh (1927)				
Modern air travelers (1996)				

B. Using the information from the chart, answer the following questions.

1. What was the fastest form of long-distance transportation in the fifteenth and

    sixteenth centuries? _____

2. What is the fastest form of long-distance transportation today?

    _____

3. In what century did the greatest advances occur in traveling long distances in a
    short time? What method of travel made these advances possible?

    _____

4. Which of the forms of transportation listed on the chart will probably continue to

    get faster as technology advances? _____

5. Use a calculator to figure out how many miles per hour Charles Lindbergh covered
    on his trip from New York to Paris. Compare that figure with a flight in 1996 from
    New York to London by supersonic jet. List the figures below.

    Charles Lindbergh's miles per hour _____

    Modern air travelers' miles per hour _____

C. Complete the following regarding future travel and tools.

1. The people you studied traveled over land, across water, and through the air. Today
    people are exploring other parts of our world in various ways. List some places that
    are still being explored.

    _____

    _____

2. How do you think people will travel to these places that are still being explored?

    _____

    _____

3. What do you think people will find in these new places? For example, what do
    you think explorers would find in space that would improve life on Earth?

    _____

    _____

Lesson 6

Explore! Explore!

As you read about reasons for the European era of exploration, think about how those reasons were related to geography.

"Explore! Explore!" might have been the words government officials spoke to their people as they encouraged exploratory missions in the 1500s. But many things actually pushed and pulled sixteenth-century Europeans from their homeland to sail uncharted seas.

Some Europeans set sail for religious reasons. In the sixteenth century, European nations were Christian. Many kings and queens launched expeditions to take Christianity to people in faraway lands. These rulers believed that it was their duty to spread their faith across the world.

Another factor that encouraged exploration was the fact that farmland in sixteenth-century Europe was becoming scarce. In Spain, for example, overgrazing by sheep had ruined much of the farmland. Some people hoped to find new, unspoiled farmland and jobs in another part of the world.

Trade was also a reason for exploring new lands. Sixteenth-century Europeans used lots of spices to enhance the taste of their meats. The geography of Europe at that time was not suitable for growing pepper, cloves, and other spices. Therefore, the Europeans got their spices from Asian countries such as India and China. However, the Europeans did not trade directly with the Asians because merchants from the Middle East controlled the trade routes. Explorers were sent east and west to find new trade routes that would allow them to avoid the Middle East merchants serving as middlemen.

Maps are important tools for exploration.

Precious metals, fame, and fortune also prompted Europeans to explore new lands. In the sixteenth century, Europeans needed more silver and gold to make coins to pay for the goods they bought. The search for these precious metals lured many sailors to other lands. The desire for fame and fortune came with the search for silver and gold. Imagine the riches and glory that would belong to those who found silver and gold and a new trade route to the East!

Lastly, technology made it possible for sixteenth-century Europeans to explore new routes and lands. Ships, maps, and navigational instruments enabled them to go farther than people had ever gone before.

The Age of Exploration in the fifteenth and sixteenth centuries was largely prompted by overused land that was unsuitable for farming and depleted of its resources. These were some of the factors that pushed and pulled people to explore new lands. Today people are still encouraged to explore new places and technology for similar and different reasons.

Explorers were prompted to sail uncharted seas for various reasons.

Lesson 6

MAP SKILLS Using Latitude and Longitude to Find a Location

Latitude and longitude lines are horizontal and vertical lines drawn on a map to help you find and describe places on Earth. Lines of latitude are horizontal lines that circle the globe from east to west. Lines of longitude are vertical lines that extend from north to south on the globe. Lines of latitude and longitude provide the absolute location of a place.

The Voyage of Christopher Columbus, 1492

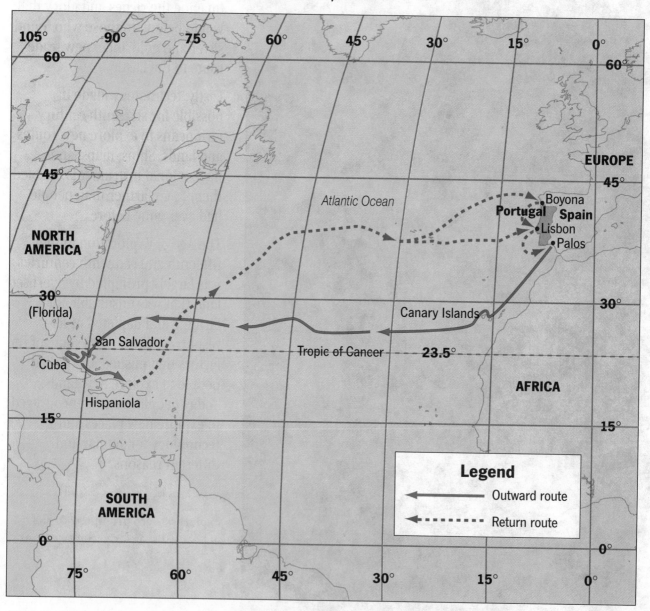

A. Look at the map of Columbus' travels in 1492 and answer the following questions.

1. Columbus left Palos, Spain, on August 3, 1492. What is the approximate latitude of Palos, Spain?

   _____

2. What is the approximate longitude of Palos, Spain? _____

3. What are the approximate latitude and longitude of the Canary Islands, where Columbus stopped in August 1492?

   _____   _____

4. If Columbus was at 30°N latitude and 60°W longitude, where would he be?

   _____

5. At about what latitude was Columbus when he landed on San Salvador?

   _____

6. Just south of San Salvador is a line of latitude with a special name. It is the northern boundary of the tropical region. What is this line of latitude called?

   _____

7. If Columbus had landed at 30°N latitude on the North American continent, where would he have landed?

   _____

B. Looking at the map, number the following events of Columbus' journey in the correct order. Beside each different place where Columbus stopped, write its latitude and longitude.

Sequence	Events	Latitude and Longitude
	Columbus lands at San Salvador Island.	
	The Santa Maria goes aground on a reef.	
	Two ships return to Palos.	
	The Niña and Pinta begin their return voyage.	
	Columbus reaches the Canary Islands.	
	Columbus sails three ships out of port Palos, Spain.	

Lesson 6

ACTIVITY   Find out why people explore new places.

Exploring New Places

The **BIG** Geographic Question

What motivates people to explore and to move to new places?

From the article you learned some reasons the sixteenth century became the great age of European exploration. The map skills lesson showed how to locate places around the world using lines of latitude and longitude. Now think about what would have made you want to explore if you had lived in the 1500s. What would make you want to explore now?

A. Push/pull factors are reasons that make one decide to travel to a new place, or stay at home. **Push factors** are influences at home that encourage people to move away. **Pull factors** are influences in a new place that encourage people to move there.

1. Think about what you read in the article. Make a list of sixteenth century push/pull factors.

Push Factors	Pull Factors
_____	_____
_____	_____
_____	_____
_____	_____

2. Select one character to represent from the following list of people who were interested in exploration in the sixteenth century. Circle your choice.

king/queen   merchant   mapmaker

European citizen   explorer   sailor

36

B. Using the idea of push/pull factors, think about why you, a sixteenth-century individual, want to leave your home and explore a faraway, unknown place. Or consider why you want to stay home and support a trip made by an explorer. Think about how you will be able to make or support a trip. Do you have a marketable talent or skill? Do you have resources for a voyage of exploration? Write a couple of sentences that describe your motivation and your talent.

_____

_____

_____

C. The great age of European discovery took place almost 500 years ago. Are there still places to explore? If so, where?

_____

_____

D. Think about these places to explore today.

1. Make a list of the push/pull factors that might be related to this present-day exploration.

Push Factors	Pull Factors
_____	_____
_____	_____

2. Select one of the following characters who might be interested in exploration today. Circle your choice.

The President   merchant   scientist

American citizen   explorer   geographer

3. Think about how you will be able to participate in or support the exploration. Do you have a marketable talent or skill? Do you have resources that would be helpful to exploration? Write a couple of sentences that describe your motivation and talent or resources.

_____

_____

E. Discuss your sentences with a classmate or friend and get feedback from them. Write on a separate sheet of paper a rap song, skit, poem, or story about your exploration. Present your work to the class.

Lesson 7

Making a Living in a New Land

As you read about the colonies different European cultures established in North America, think about how each culture used and adapted to the new land.

The Spanish had been in the Americas since the late 1400s. There they had conquered the Aztecs and Incas and had taken the silver and gold from their temples. The Spanish created large ranches to maintain horses and mules for their mining work. The Spanish also built missions in areas such as present-day California, Texas, and Florida because they thought they should convert the Native Americans living in North America to Christianity. Spanish settlements were established on both the southeastern and southwestern coasts of North America.

The first permanent English colony was established in 1607 in Jamestown, Virginia. The early settlers of Jamestown were not prepared for the hard work of earning a living on land that was wilderness. Starvation, disease, and conflict with Native Americans almost ruined the colony. Then the settlers learned to raise tobacco from the Native Americans. Virginia soon grew wealthy as a colony of large farms, called **plantations,** that grew tobacco, corn, and other crops which the settlers were able to sell in Europe.

Shortly after the English, French settlers arrived in North America in 1608. They claimed the area from the St. Lawrence River in present-day Canada to present-day Louisiana, including land along the Mississippi River. The French settlers made a living by hunting the small animals that lived in the forests. They trapped the animals for furs and started a very active fur trade because of the Europeans' demand for pelts. Pelts are the skins and fur of animals that are used for making garments.

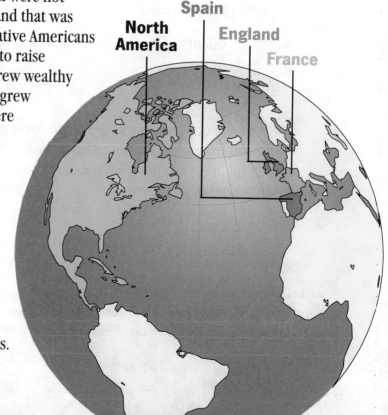

North America   Spain   England   France

In 1620 the Pilgrims established Plymouth Colony in the area that is now Massachusetts. The Pilgrims had separated from the Church of England and wanted to raise their families where they could live according to their own religious beliefs. They had to learn to make a living on thin, rocky soil in a climate with long, cold winters. When they were not successful at fishing, they turned to farming and fur trading and began shipbuilding.

When Europeans first arrived in North America, they brought some of their own customs, traditions, and ways of life to North America. However, in large part, they found ways to adapt to the land of the new country.

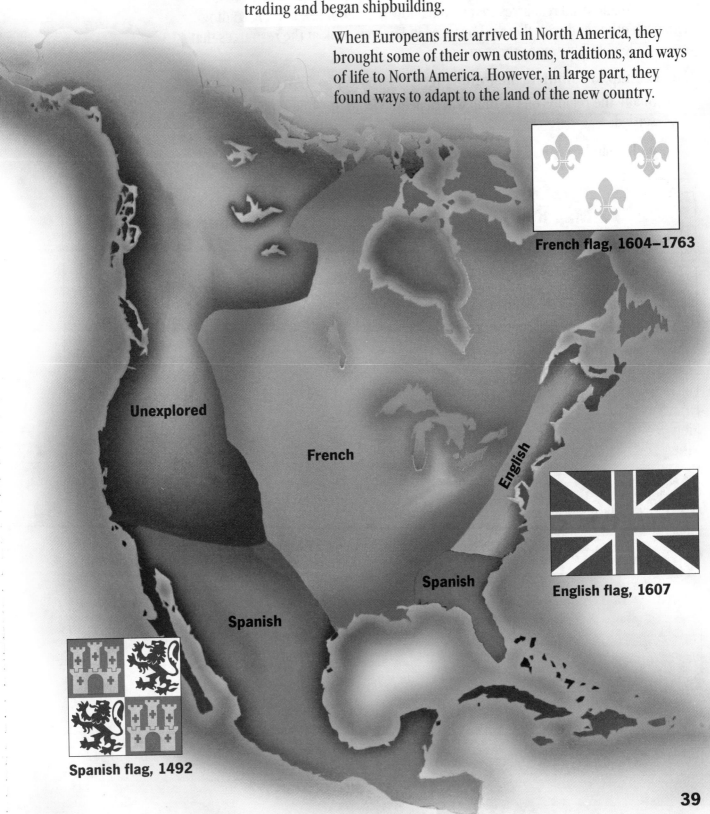

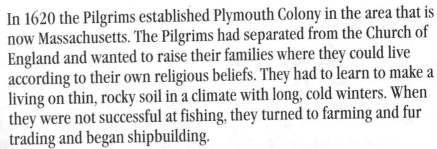

French flag, 1604–1763

Unexplored

French

English

Spanish

English flag, 1607

Spanish

Spanish

Spanish flag, 1492

MAP SKILLS
Using a Map to Look at Resources of Early Settlements

The Spanish, English, and French settled areas of the present-day United States. They found and used resources that would allow them to make a living trading with each other and with those who stayed in Europe. Look at the resources that made trade possible among these groups.

Use the map to complete the following.

1. Shade the regions on the map where people from the following countries settled.

   a. England   **b.** France   **c.** Spain

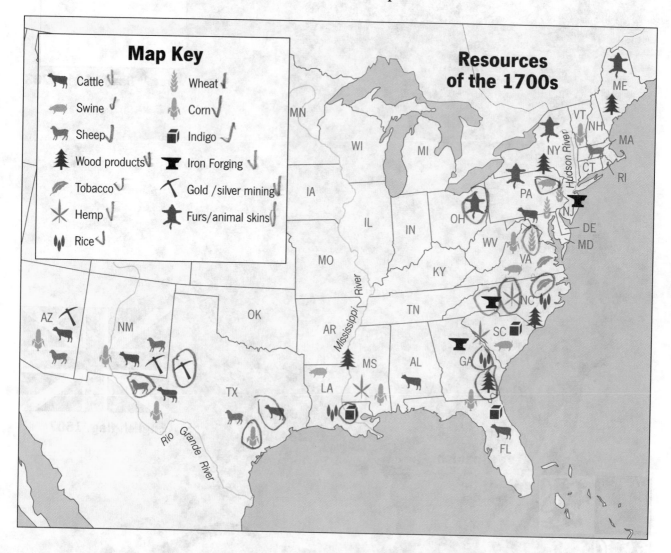

Map Key

- 🐄 Cattle
- 🐖 Swine
- 🐑 Sheep
- 🌲 Wood products
- 🍃 Tobacco
- ✳ Hemp
- 🌾 Rice
- 🌾 Wheat
- 🌽 Corn
- ▯ Indigo
- ⚒ Iron Forging
- ⚒ Gold /silver mining
- 🐾 Furs/animal skins

Resources of the 1700s

2. Name two farm products raised in the 1700s by settlers from each of the following countries.

    a. England   _____   _____

    b. France   _____   _____

    c. Spain   _____   _____

3. Some of the resources the settlers found did not have to be grown, but still came from the land. These types of resources are raw materials. Name two kinds of raw materials found in the 1700s by settlers from each of the following countries.

    a. England   _____   _____

    b. France   _____   _____

    c. Spain   _____   _____

4. List below each of the products you named above and how Europeans and Native Americans might have used these farm products.

Farm Products	European Uses	Native American Uses

5. Which group of early colonists relied most heavily on farming? _____

6. Which group of early settlers relied most heavily on the land's raw materials? _____

7. Do you think the Spanish, French, and English settlers used more of the products they raised in North America or traded more with their native countries? Explain your answer.

_____

_____

Lesson 7

ACTIVITY
Find out about some early North American settlement types and locations.

Building a Settlement

The **BIG** Geographic Question
What types of settlements did groups from European nations create in North America and where were they?

From the article you learned that the English, the French, and the Spanish came to North America for different reasons and settled in different regions. In the map skills lesson you looked at some of the products the Europeans raised and how these products were used. Decide where you would have settled in North America in the 1700s.

A. Use the Almanac to identify each settlement's purpose and features of the land areas where they originated and were settled. Complete the chart below.

European Group	Type of Settlement	Purpose	Land and Climate Features in Europe	Land and Climate Features in North America
English and French				
French				
Spanish				

B. Select one of the following to design: a Spanish mission in San Antonio, Texas; a French trading post in New Orleans, Louisiana; or an English fort in Philadelphia, Pennsylvania. Write your choice on the line below.

_____

C. Look at the map on page 40. Mark the location of your settlement choice.

D. To help plan your settlement, find out about the following physical features of the area where it is located. Look in the Almanac for help with this information.

1. climate _____

2. vegetation _____

3. waterways _____

E. Draw the design of your settlement choice in the space below. List off to the side of your drawing the area's natural resources that you used in planning your settlement's construction. On the lines below, explain how the North American terrain and the settlement's future dealings with the mother country influenced its location and affected your design.

Explanation of Settlement: _____

_____

CLOSE ENCOUNTERS

As you read about the first contacts between Native Americans and Europeans, think about how their encounters affected their cultural surroundings.

"Men of strange appearance have come across the great water." These were the words a Native American prophet told his people in the 1500s.

The first contact between Native Americans and British settlers was in the northeastern part of the United States. The settlers learned from the Native Americans the best ways to use the land for growing crops such as beans, maize, squash, and tobacco. French fur traders of the east learned new hunting skills from Native Americans. In return, Native Americans traded furs for European manufactured goods.

In the southern Great Plains, Spaniards introduced European crops such as apples, peaches, and pears to the Pueblo people. In return, the Pueblo shared their different farming techniques. The Spaniards also brought sheep, goats, cows, oxen, pigs, chickens, and horses to the new land. These animals became very useful to Native Americans.

The introduction of horses in the Great Plains dramatically changed the lives of the Native Americans who lived there. Riding horses, these Plains dwellers could leave their villages to follow the buffalo for hundreds of miles. Large tepees,

Glass beads, scissors, rings, and clay pipes were some of the trade goods used to buy furs and other items from the Native Americans in the late 1600s.

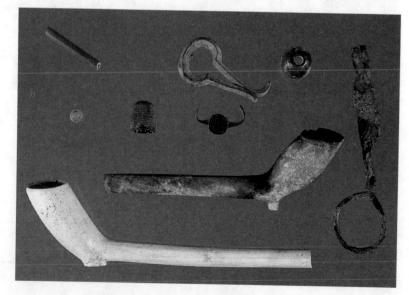

44

Cloth, metal tools, guns, and ammunition were among the most favored items Europeans traded to Native Americans.

which could be easily transported by horses, replaced the more permanent earth lodges.

As more and more Europeans settled the land and cleared it for farming, animals prized for their pelts moved farther and farther away. Eventually, the beaver habitats around the Appalachians and the Great Lakes were almost destroyed, forcing trappers and settlers westward.

As contact between European settlers and Native Americans continued, they exchanged more ideas, goods, and information. However, opinions on the value and use of the land differed. Native Americans believed that all people should share the land and its resources. They did not view land as a property to own. Representatives of the British, French, and Spanish governments wanted to claim large areas for their rulers. British farmers and Spanish livestock owners wanted to fence in areas of their own. Even if French fur trappers did not claim land, they wanted to have exclusive rights to trap in certain areas. The conflicts that arose between the Europeans and the Native Americans contributed to changes in the physical environment around them.

Lesson 8

MAP SKILLS Using a Topographical Map to View Landforms

Mapmakers use many techniques to show differences in elevation, or height of Earth's surface. Colors can be used to show the various levels of the land.

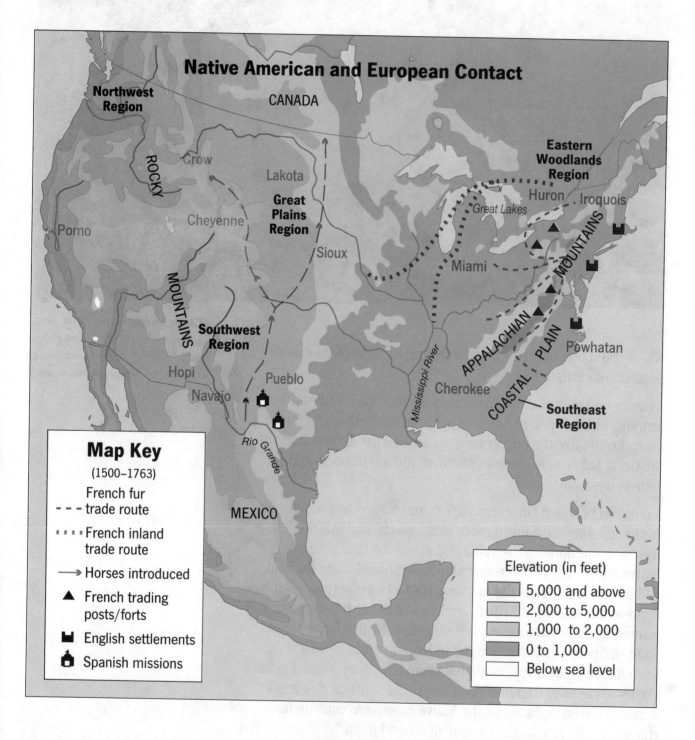

Native American and European Contact

Map Key
(1500–1763)

- - - French fur trade route

···· French inland trade route

→ Horses introduced

▲ French trading posts/forts

⌂ English settlements

⌂ Spanish missions

Elevation (in feet)

5,000 and above
2,000 to 5,000
1,000 to 2,000
0 to 1,000
Below sea level

A. Look at the map and its key. Find the elevations in the following regions of the United States.

1. Rocky Mountains _____   3. Coastal Plain _____

2. Great Plains _____   4. Central Lowlands _____

B. Use the map to help you answer the following questions.

1. French explorers were mainly interested in fur pelts rather than settling land.

   a. Through what areas did the French trade routes extend?

   _____

   b. What geographical features might have presented difficulties for the French

   traders? Why? _____

   _____

   c. Which Native American tribes might have interacted with the fur traders?

   _____

2. The English settlers were interested mainly in raw materials and food to send back to England.

   a. Which area included most of the English territory in the United States?

   _____

   b. Which Native American tribes might have taught English settlers how to

   plant and grow crops? _____

   _____

3. In what region did the Europeans introduce horses to Native Americans and why do you think horses became so valued in this region of the country?

   _____

   _____

   _____

4. Which Native American tribes might have interacted with the Spanish missionaries?

   _____

   _____

Lesson 8

ACTIVITY
Learn about the different views on how land should be used.

Debating for Land

The **BIG** Geographic Question

How should land be used, and who should own it?

From the article you learned about conflicting Native American and European ideas about using land. The map skills lesson showed you various Native American tribes and physical features of the regions in which they lived when contact was first made with Europeans. Now conduct a debate as you negotiate, or bargain, for land.

A. Using information in the article, name four key geographic regions in North America at the time of early European settlements.

1. _____   3. _____

2. _____   4. _____

B. Choose one of the regions that you listed and answer the following questions about that region.

1. Which Native American tribes lived in this region?

   _____

   _____

2. What were the major resources in the region?

   _____

   _____

3. How did the Native Americans who lived in that region use the land?

   _____

   _____

48

C. The Native Americans, English farmers, Spanish missionaries, French traders, and representatives of European governments had different ideas about how to use North America's land and resources. On the chart below, write each group's ideas about how to use the land.

Group	How They Wanted to Use the Land

D. Evaluate each group's ideas about how to use the land and its resources and whether land should or should not be owned.

1. Choose a point of view to represent.

2. Write notes below to support your point of view and to prepare for a debate.

_____

_____

_____

_____

_____

_____

_____

3. Debate the issue of land use and ownership with a classmate who chose to represent a different point of view.

Lesson 9

New York— A "Capital" City

As you read about New York City, think about how its geography influenced its selection as the first United States capital.

After the signing of the Declaration of Independence in 1776, Congress met in nine different cities trying to select a capital. New York was one of the cities that wanted to be chosen as the nation's capital.

New York, one of the original colonies, was a major port and center of trade. The people of New York wooed Congress with many promises of support. On January 11, 1785, the Continental Congress moved to New York City and stayed there for five years. During that time, the states approved the Constitution, and George Washington was inaugurated as the nation's first president in New York.

The choice of any existing city as the nation's capital wasn't easy. People disagreed on how the new nation should be run. Congress couldn't be sure that the state would defend it in case of attack. Finally, Congress decided to choose a new city outside the bounds of any state. Maryland and Virginia donated land along the Potomac River. The city of Washington, D.C., was founded as the nation's new capital.

A map of New York City, 1767

THE RATZER MAP
OF
NEW YORK CITY
1767.

When Congress moved away, many New Yorkers were upset because they had made great efforts to please Congress. However, New Yorkers soon forgot their loss. With its excellent natural harbor, New York increased its trade with foreign countries. Businesses grew and needed more workers. Thousands of European immigrants poured into the city.

The original city of New York occupied the southern tip of Manhattan Island. By 1898 bridges and roads had improved transportation greatly. Then Manhattan joined with Brooklyn, Queens, the Bronx, and Staten Island to form Greater New York city, increasing the area tenfold. Its 3,500,000 people made it the second largest city in the world. Only London was larger.

During the 1900s New York City continued to grow. Giant companies made their homes there. Great skyscrapers, such as the Empire State Building, turned the streets into "stone canyons." Today, New York City is one of the world's most important centers of business, trade, finance, and the arts. The city that was once the nation's capital is still one of the nation's "capital" cities.

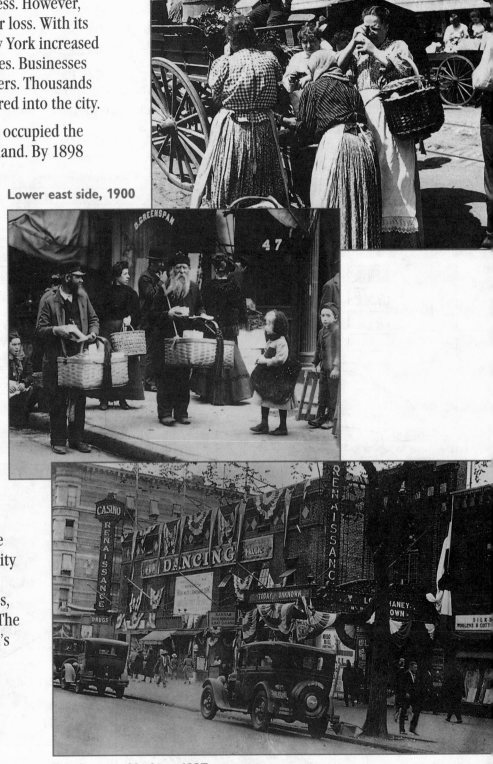

Hester Street, 1898

Lower east side, 1900

7th Avenue, Harlem, 1927

MAP SKILLS
Using Large- and Small-Scale Maps

The area and amount of information that a map can show depends on its scale. On a **small-scale map** you get general information about a larger area. On a **large-scale map** you get much more detailed information about a small area.

Locator map

Map Key

POPULATION DENSITY
Number of persons per square mile

	more than 250
	100 to 250
	50 to 100
	Less than 50

CANADA

NEW YORK

VT

NH

Lake Ontario

Rochester

Buffalo

Syracuse

Utica

Lake Erie

Finger Lakes

APPALACHIAN MOUNTAINS

Albany

MA

CT

RI

PA

Boroughs of New York

N
W E
S

New York

The Bronx

Manhattan

Queens

New Jersey

Brooklyn

Staten Island

New York City

Long Island

NJ

State map

City map

52

A. Use the New York state map to answer the following questions.

1. Where are the areas of greatest population density in the state?

_____

2. How would you account for the low population density in the light yellow?

_____

3. What geographic feature might account for the population density represented by a strip of orange through the middle of the state?

_____

B. Use the map of New York City to answer the following questions.

1. What are the five boroughs of New York City?

_____

2. Which borough is the southernmost borough?

_____

3. Which boroughs are located on Long Island?

_____

C. Compare the two maps and answer the following questions.

1. From which map would you get more detailed information, the city map or the state map?

_____

2. Which is the large-scale map?

_____

3. Which is the small-scale map?

_____

4. Which map would you use to compare the populations of several cities?

_____

5. On which map could you find more information about a specific area?

_____

Lesson 9

ACTIVITY   Create a visual display representing
New York City's population change over time.

New York City—Then and Now

The **BIG**
Geographic Question

How can we show changes in population over time?

From the article you learned about the role of New York City as the first capital of our nation, a port of immigration and trade, and a center of business and culture. From the map skills lesson you learned how the city's population is distributed. Now discover how New York City's population has changed over time.

A. Look at the following table of New York City's population from 1750 through 1990.

  1. Using the table, draw a bar graph showing the population for each year shown. The year 1750 has already been done for you.

  2. Give your graph a title.   Title _____

Year	Population
1750**	25,000
1800	60,000
1850	696,000
1900	3,400,000
1950	7,900,000
1990*	7,300,000

Figures are rounded.

*As of last U.S. Census

**Estimated

Millions — (y-axis: 1, 2, 3, 4, 5, 6, 7, 8)

Time (years) — (x-axis: 1750, 1800, 1850, 1900, 1950, 1990)

54

B. Use your graph and information from the article to answer the following questions.

1. What were some of the factors that probably led to the increase in New York City's population:

   a. from 1750 to 1800? _____

   _____

   b. from 1800 to 1850? _____

   _____

   c. from 1850 to 1900? _____

   _____

   d. from 1900 to 1950? _____

   _____

2. Between 1950 and 1990, the population changed as follows:

1950	7,900,000
1960	7,800,000
1970	7,900,000
1980	7,100,000
1990	7,300,000

   a. Why do you think the population dropped between 1970 and 1980?

   _____

   _____

   b. What might you expect the population to be at the next census in the year 2000? Explain your answer.

   _____

   _____

C. Create a visual aid, such as a bar graph, pie chart, or table to show New York City's ethnic population mix in 1990.

1. Look in the Almanac to find the information you need.
2. Organize your information by using the percentages to divide the 1990 bar of your graph on page 54 into sections.
3. Create your visual aid and label each section with the ethnic group it represents.

Lesson 10

A Battle for Survival

As you read about Valley Forge,
think about the role geography played in making
it an important historical site even though
no military battle was fought there.

Valley Forge was located between Philadelphia and York. The forge itself, located on Valley Creek, had once processed iron. The iron was used to make many items, including military supplies. After their victory at Brandywine, British soldiers raided Valley Forge. They took all the supplies and then burned the forge to the ground.

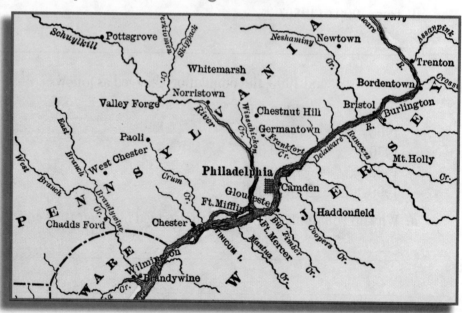

Enlarged view of a historical map of the layout of Valley Forge.

Despite its bleakness, General George Washington chose this site for his winter camp in 1771. It was more easily defended than other places near Philadelphia because of the Schuylkill River and the steepness of the valley walls. Because the Schuylkill River flows by Valley Forge on its way to Philadelphia, where it empties into the Delaware River, Washington's scouts could keep track of General Howe's army. They were spending the winter in the warmth of Philadelphia. Valley Forge also placed Washington's army between Howe and the Congress at York.

During the summer of 1777 the Continental Army, led by Washington, won several victories at Trenton and Princeton in New Jersey. But in the fall, British General William Howe defeated Washington's army at Brandywine Creek and Germantown in Pennsylvania.

On December 19, 1777, Washington led more than 11,000 poorly trained troops down Gulph Road to this winter camp. The men were tired and disappointed over their recent losses, and they lacked food, clothing, and blankets. Soon after their arrival, Washington wrote the Congress that "2,898 men are unfit for duty because they are barefoot or otherwise naked."

Washington crossing the Delaware

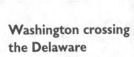

Albigence Waldo, *a doctor serving with the army, wrote:*
"I am sick ... poor food ... cold weather ... fatigue ... nasty clothes ...
There comes a bowl of beef soup—full of burnt leaves and dirt...."
Yet Waldo was amazed at the men. "They show a spirit ... not to be
expected from so young Troops."

Matters got worse. The quartermaster, an officer responsible for supplying the army, was poorly organized. Soldiers were sent into the surrounding countryside to find cattle and other food. Local people preferred selling their goods to the British because Continental currency had little value. Twelve men lived in each hut, where they had little water with which to wash. After months in these cramped quarters, many soldiers had fallen ill.

By February of 1778 the situation seemed hopeless. Only five thousand of the troops were fit for action. Nearly one fourth had died from malnutrition. Then, at this lowest point, the tide turned. New recruits began to arrive. A new quartermaster was appointed, and supplies increased. Baron von Steuben, a Prussian officer, arrived to train the men. Although he spoke no English, his drill sergeant manner worked. The soldiers grew stronger and more disciplined.

The only battle fought at Valley Forge was the battle against suffering and hardship, against starvation and death. The Continental Army that left Valley Forge on June 19, 1778, was very different from the one that had arrived. Their experiences at Valley Forge had produced an army that never stopped until they defeated the British at Yorktown.

Lesson 10

MAP SKILLS
Using a Map to Find Location in Latitude and Longitude

Lines of latitude and longitude are drawn on a map to help us identify a place's specific location on the Earth's surface. These lines are measured in **degrees,** indicated by the ° symbol. Lines of latitude and longitude are expressed in numbers, often called coordinates.

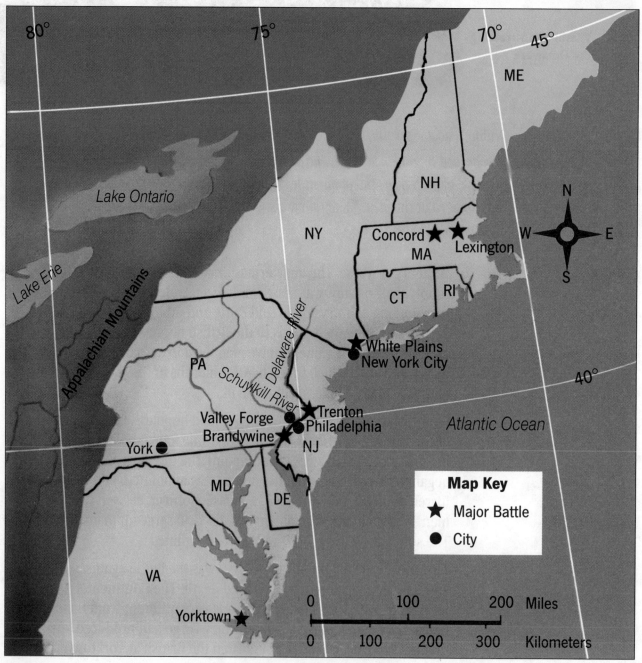

80° 75° 70° 45°

ME

Lake Ontario

NH

NY

Concord ★ ★

Lexington

MA

N

W E

S

Lake Erie

CT RI

Appalachian Mountains

Delaware River

White Plains
New York City

40°

PA

Schuylkill River

Valley Forge ★ Trenton

Brandywine Philadelphia

Atlantic Ocean

York ⬤

NJ

Map Key

★ Major Battle

⬤ City

MD

DE

VA

Yorktown ★

0   100   200   Miles

0   100   200   300   Kilometers

A. Use the map to answer the following questions about specific location.

1. What city lies at a latitude of 41°N and a longitude of 74°W?

   _____

2. About how many degrees of latitude are there between Philadelphia and New

   York City? _____

   _____

3. About how many degrees of longitude are there between White Plains, New York

   and York, Pennsylvania? _____

4. The Schuylkill River lies between what lines of latitude? _____

B. Look at the entire area that the map shows and answer the following questions.

1. About how many degrees of longitude does the map cover? _____

2. About how many degrees of latitude does the map cover? _____

C. Using the map, identify the places described below and answer the following questions.

1. If point A is located at 42°N and 75°W and point B is located at 40°N and 74°W, in what direction would you travel to get from point A to point B?

   _____

2. Is there a city at 39°N and 74°W? If so, what city is it?

   _____

3. About how many degrees of latitude are there between Yorktown, Virginia, and Philadelphia, Pennsylvania?

   _____

4. What are the latitude and longitude coordinates for Lexington, Massachusetts?

   _____

5. What are the latitude and longitude coordinates for York, Pennsylvania?

   _____

Lesson 10

ACTIVITY

Explore the effects of physical geography on various Revolutionary War battles.

Geography in Battle

The **BIG** Geographic Question

How did physical features affect the outcome of Revolutionary War battles?

From the article you learned about conditions at Valley Forge and why Washington chose that location for his army's winter camp. From the map skills lesson you learned about the location of some Revolutionary War battles. Now find out about three other Revolutionary War battles.

A. Use the Almanac to complete the following.

1. Identify and list the physical features surrounding the following battle sites.

   Yorktown: _____

   Lexington/Concord: _____

   Trenton: _____

2. In which battles was weather an influencing factor? Explain why. _____

   _____

   _____

B. Use the Almanac to read more about the battle sites listed above. Make notes in the chart below.

Yorktown	Lexington/Concord	Trenton

60

C. **Look back at the physical features you listed for each battle site.**

   1. Suppose you were the defending army at each battle location.

   2. Put a + above the features that you think were positive location factors.

   3. Put a – above the features that you think were negative location factors.

D. **Compare your list of positive and negative factors to your notes from the Almanac.**

   1. Did you come up with the same physical features and locational factors?

   _____

   2. Were you accurate in identifying which physical features would be helpful to you as a defending army? Explain your answer. _____

   _____

   _____

   _____

   _____

E. **Describe how the geographic setting of each battle site affected the outcome of the battle.**

   _____

   _____

   _____

   _____

   _____

   _____

   _____

   _____

   _____

Lesson 11

Westward Wagons

As you read about early pioneers who explored the west, think about how the land might have looked if you were traveling in an open wagon!

The United States began as thirteen colonies lying east of the Appalachian Mountains, extending along the Atlantic Ocean from Maine to Georgia. By the end of the Revolutionary War in 1783, farmers and ranchers were settling the fertile land of the Mississippi Valley. The more adventuresome—fur traders, explorers, and missionaries—forged on into the vast, unexplored lands west of the Mississippi. Beyond the dry, high plains known as the "Great American Desert," the Rocky Mountains rose as a forbidding wall, defying people to pass.

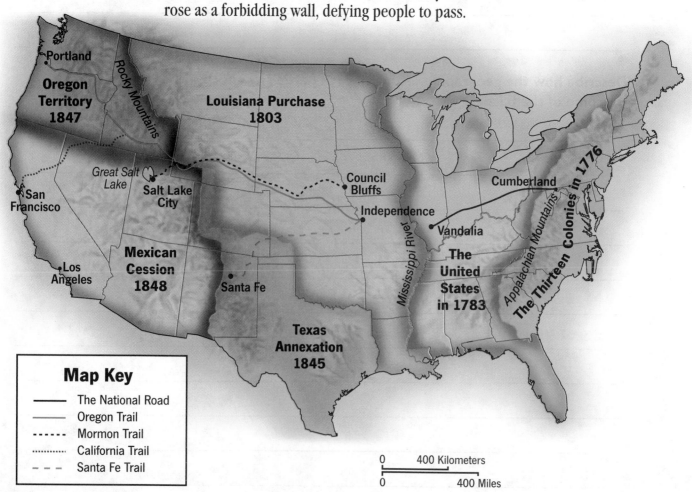

Portland

Oregon Territory 1847

Rocky Mountains

Louisiana Purchase 1803

Great Salt Lake

Salt Lake City

San Francisco

Council Bluffs

Independence

Cumberland

The Thirteen Colonies in 1776

Vandalia

Mexican Cession 1848

Los Angeles

Santa Fe

The United States in 1783

Appalachian Mountains

Mississippi River

Texas Annexation 1845

Map Key
——— The National Road
——— Oregon Trail
- - - - Mormon Trail
·········· California Trail
– – – Santa Fe Trail

0   400 Kilometers
0   400 Miles

Explorers came back from these western regions with stories of rich soil, a farmer's paradise, a land flowing with milk and honey. Worried about low crop prices and failing soil, many farmers and ranchers in the East were tempted by the promise of more and better land. They were also attracted by waterways that allowed for easy transportation of their goods. Although they knew the journey westward was dangerous, the stories of rich, fertile soil pulled them westward. Publisher Horace Greeley said that the first migration of more than 1,000 people to Oregon ". . . wears an aspect of insanity." But in 1843, that first wagon train succeeded.

How might those brave pioneers have described their journey across rivers, mountains, and desert? Here is how excerpts from the diary of a young pioneer who went west might have read:

19 April—The wagon looks so nice, with its white cover. With everything packed away shipshape, it is a prairie schooner indeed. The oxen move very slowly so we have no trouble walking beside it.

21 May—We rose at 4 A.M. to collect buffalo droppings for the fire because there is no wood. Mother and I cooked a big breakfast of pancakes, meat, biscuits, and beans. We all drink coffee because the water tastes so bad.

15 June—We have gone for days through this sea of dry grass. Will it ever end! My brother shot a buffalo. It was so big, but we dried a lot of the meat and cleaned the hide for a blanket.

23 July—This river must be a mile wide. We floated our wagons across on rafts. Some of the cattle got caught in quicksand and drowned.

4 August—Today we rested and got supplies. The mountains look impossible to pass, but our leader says that there is a trail. We must trust him.

25 Sept.—It has been five months since we left home, but we have reached Oregon! The land is better than we expected. I miss the friends who died along the way. But we will have a cabin and rich crops. Life will be good again.

Lesson 11

MAP SKILLS

Using a Map to Interpret the Length and Difficulty of a Trip

We can use a map to figure out how far places are from one another. We can also use them to show what geographic features lie along a certain route. A map can give you some idea of what a pioneer's journey west was like.

A. Look at the map of the Oregon Trail below. Circle the following.

1. The North Platte River
2. Fort Laramie
3. Independence
4. The Snake River
5. The Rocky Mountains
6. South Pass
7. Oregon City, Oregon
8. The Great Plains

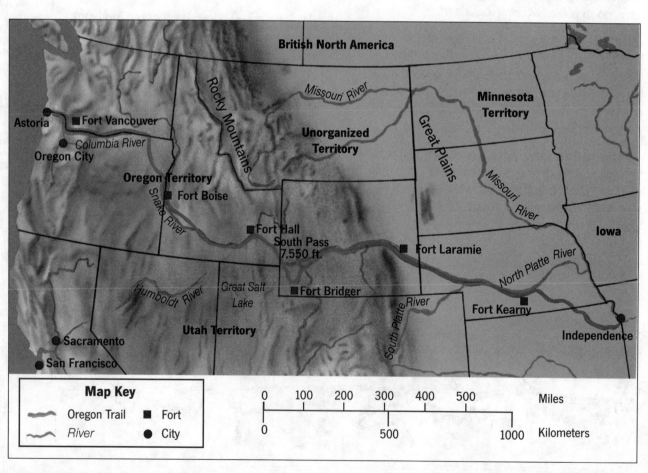

British North America

Missouri River

Rocky Mountains

Minnesota Territory

Astoria ■ Fort Vancouver

Columbia River
Oregon City

Unorganized Territory

Great Plains

Oregon Territory

■ Fort Boise

Snake River

Missouri River

Iowa

■ Fort Hall
South Pass
7,550 ft.

■ Fort Laramie

North Platte River

Humboldt River

Great Salt Lake

■ Fort Bridger

South Platte River

Fort Kearny

Utah Territory

● Sacramento

● San Francisco

Independence

Map Key

～ Oregon Trail   ■ Fort
～ *River*   ● City

0   100   200   300   400   500   Miles

0   500   1000   Kilometers

B. Use the map and the Almanac to help you answer the following questions.

1. Through what present-day states did the Oregon Trail pass? _____

_____

2. In what mountains is South Pass located? _____

   How high is South Pass? _____

3. What major rivers did the pioneers follow during their journey?

_____

_____

C. Use a piece of string, a ruler, the map, and the map scale to figure out the following times and distances.

1. What is the distance between Independence and Fort Laramie?

_____

2. How long is the Oregon Trail from Independence to Oregon City?

_____

3. If the wagon train averaged 15 miles per day, how long would it take them to make the journey from Independence to Oregon City?

_____

D. Look at the map and describe what a journey along the Oregon Trail would have been like.

1. Why do you think the trail mostly followed rivers?

_____

_____

2. What geographic features would have slowed the settlers down? What features would have given them the greatest challenge? Explain your answers.

_____

_____

_____

_____

Lesson 11

ACTIVITY

Write a diary entry describing how it might have felt to travel across the country in a covered wagon.

Dear Diary

The BIG Geographic Question

What physical challenges did the pioneers meet as they moved westward and how did they handle them?

From the article you learned what life was like for the pioneers who settled the West. In the map skills lesson you identified the physical features that the pioneers faced on the trail west. Now explore and describe how people felt and what they thought as they made their difficult journeys westward.

A. Select one of the trails shown on the map of westward expansion on page 62: the National Road, the Oregon Trail, the Mormon Trail, the California Trail, or the Santa Fe Trail. Answer the following questions. Use the Almanac to find out more about the trail you chose.

1. How many miles long was the trail?

_____

2. What mountain ranges, rivers, deserts, or other physical challenges did the people traveling it encounter?

_____

_____

3. What sources of food and water might the pioneers have had along the way?

_____

_____

4. Did the pioneers cross any land that was occupied by Native Americans? Explain your answer.

_____

_____

_____

B. Imagine that you are a member of a pioneer family traveling on the trail you selected and keeping a diary. Complete the following to help organize the information that will go in your diary entries. On a separate piece of paper, write two or three entries describing some part of your trip. Date the entries in your diary with appropriate dates.

1. Describe how you felt about leaving home.

   _____

   _____

2. Describe various physical features that you encountered during your journey westward.

   _____

   _____

3. Describe any hardships or challenges that your family faced and tell how you overcame them.

   _____

   _____

4. Describe what a typical day was like. (What did you eat? How did you cook? How many miles did you cover in a day? What did you use for fuel? What were your duties? Did you have time for play or other activities?)

   _____

   _____

5. Describe how you felt about the trip west by considering the following questions. Were you afraid or excited? Were you worried about making the journey? Did you find the conditions of the trip depressing, or were you optimistic and amazed by the things you saw?

   _____

   _____

6. Include a final entry about what it was like to reach your destination.

   _____

   _____

C. Create a mental map (or a map of how you imagined the trail to be based on a picture in your mind) of your route west.

Lesson Overview for pages 68–73: Students use written and graphic information to learn that historical expeditions west led to the development of a system for measuring and dividing land.
Cross-Curricular Connection: History — Understands the international background and consequences of the Louisiana Purchase, the War of 1812, and the Monroe Doctrine; understands the influence of American explorers

Lesson 12

Blazing the Western Trail

As you read about famous explorers who traveled west, think about how they used their findings to map out the land.

In the early 1800s the vast lands west of the Mississippi were a mystery to most people living in the United States. After the Louisiana Purchase in 1803, the West opened to groups of explorers, hunters, naturalists, and other adventurers. Over the next 75 years, more than two million square miles of previously unexplored land was explored in a series of scientific expeditions. Although the land was considered "property" to the settlers, it was not free for the taking. Native Americans inhabited the land.

Meriwether Lewis and William Clark's 3,000-mile expedition began in May 1804 near St. Louis, Missouri. They journeyed up the Missouri River in two dugout canoes and a flat-bottomed boat called a keelboat. They continued west by canoe through what is now North Dakota and Montana. The trip across the Rocky Mountains proved to be difficult. Horses lost their footing along the narrow mountain paths and fell to their deaths, taking precious equipment and supplies with them. Nonetheless, the explorers succeeded in crossing the mountains and traveled along the treacherous falls and rapids of the Columbia, Snake, and Clearwater Rivers to the Pacific coast. On the return trip, Lewis and Clark separated and followed different paths back to St. Louis. The detailed maps they brought back served to guide traders and trappers throughout the 1800s.

Explorers, such as Zebulon Pike, who traveled west enjoyed this panoramic view of Pikes Peak.

68
National Geography Standards: 4, 13, 14, 16
Basic Geography Skills: Acquiring and analyzing geographic information
Geographic Theme: Movement — Humans interacting on Earth

The Lewis and Clark expedition was the first of many such journeys. Another American explorer, Zebulon Montgomery Pike, traveled through the southwest between 1805 and 1806. While following the Arkansas River, he sighted a faraway peak. This peak, the first of the Rocky Mountain peaks that can be seen by travelers approaching from the east, became known as Pikes Peak. Pike tried to climb the 14,110-foot peak, but had to turn back when supplies ran low.

As settlers began to move farther west in the 1830s, it became necessary to map and survey the land. Boundaries and transportation routes needed to be marked. Land suitable for farming and mining needed to be identified. By using a system of horizontal and vertical measurements, the shape and position of a piece of land can be calculated. The Army Corps of Topographical Engineers began to map and analyze the land of the West.

From 1842 to 1844 John C. Frémont, a well-known Corps engineer, explored much of the land between the Rocky Mountains and the Pacific Ocean. From present-day Kansas City, Frémont's first expedition followed the Big Blue and North Platte rivers, charting the Oregon Trail. He completed the mapping of the Oregon Trail during his second expedition in 1845 and 1846. Frémont helped produce the first scientific map of the western United States.

In 1869 the United States government began financing a series of mapping expeditions known as the Great Surveys. John Wesley Powell and his party led teams of scientists, mapmakers, and photographers on the first expedition to the Grand Canyon and the surrounding region. Powell, like many other explorers, stressed the importance of using the land and its resources carefully. The Great Surveys became the basis for all future development of the West.

John C. Frémont was often called the "Pathfinder" because of his expeditions West.

Lesson 12

MAP SKILLS
Using a Map to Trace Expeditioners' Routes and Divide the Land

Maps can show the movement of people across an area of land. The map below uses colored arrows and different types of lines, for example, a solid versus a dotted line, to indicate different routes. The physical features of the routes are included on the map in order to show the geography encountered along the way.

A. Look at the map of expeditioners' routes below. Circle the following.

1. Grand Canyon   **2.** Snake River   **3.** Pikes Peak   **4.** Platte River

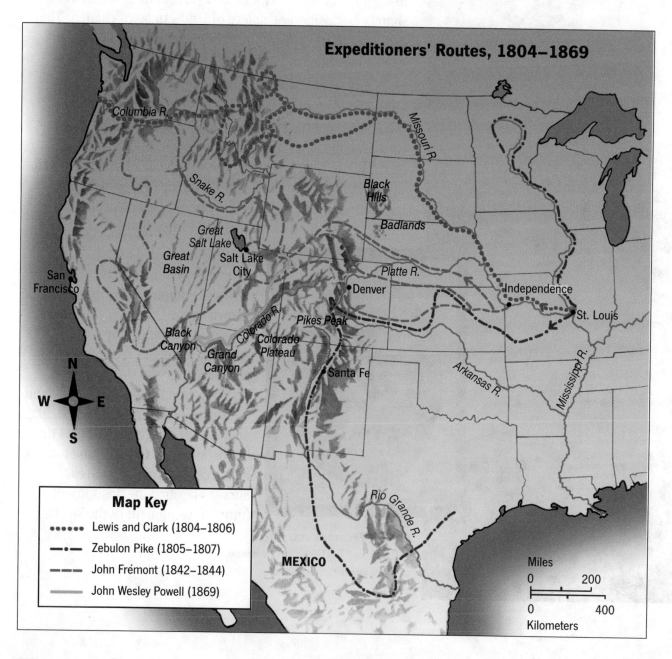

Expeditioners' Routes, 1804–1869

Map Key
- •••••• Lewis and Clark (1804–1806)
- —•—•— Zebulon Pike (1805–1807)
- — — — John Frémont (1842–1844)
- ——— John Wesley Powell (1869)

B. The map shows the routes of four famous expeditioners who went West. Using the map, answer the following questions.

1. Which explorer circled the land around the Great Basin?

_____

2. Which explorer traveled up the Arkansas River into the mountains, and then crossed the Rio Grande twice?

_____

3. Choose one of the explorers listed on the map key. Write a description of the route he followed. What kinds of landforms did the explorer encounter?

_____

_____

C. Physical features can be used as boundaries when dividing land. Suppose you were a settler traveling westward in the mid-1800s.

1. Choose a destination for your trip.

_____

2. What route would you follow to arrive at your destination?

_____

3. What kind of land features might you find along the way?

_____

_____

4. How might you use the land where you settle?

_____

_____

5. Your wagon train has a total of 30 wagons, each containing one family. Devise a method for dividing the land where you will settle. What boundaries will surround each family's plot of land? How will you decide what boundaries to use?

_____

_____

_____

Lesson 12

 **ACTIVITY** Create a work of art to represent a historical expedition.

Recreating History in Art

The BIG Geographic Question — What was the scenery and daily life of the people in the American West like?

From the article you learned about several explorers of the West. The map skills lesson showed you the routes of their expeditions. Now imagine you are an artist who has been hired to go along on one of these trips and illustrate what the explorers saw on their journeys.

A. Choose one of the expeditions described in the article and charted in the map skills lesson and write the explorer's name below.

_____

B. Answer the following questions about the land that your expedition party will explore. Look at pages 70–71 and in the Almanac for help with this information.

1. Describe the route you will travel. _____

_____

2. Record information about the following:

   a. waterways _____

   b. mountains _____

   c. other landforms _____

3. Describe what you think the climate would be like in the areas as you traveled through them. Use what you know about the landforms and the time of year in which you are traveling.

_____

_____

_____

C. Sketch a map below of the route you traveled. Be sure to indicate the landforms and water forms you listed on page 72.

D. Describe some of the living things that you see on your journey. See the Almanac information on western trails for help with this. Take notes on the chart below.

People	Animals	Vegetation

E. Use the information you have gathered to create a painting or a drawing that shows the landscape and scenery you saw. Write a brief description to go with your painting or drawing.

Lesson 13

North vs. South

As you read about the resources that the North and the South had, think about how these resources affected the outcome of the Civil War.

The North (Union) fought the South (Confederacy) in the Civil War between 1861 and 1865. The North was larger than the South, both in physical size and in population. A total of 25 northern states fought with the Union, while only 11 southern states joined the Confederacy. Almost 22 million people—including native-born whites, immigrants, free African Americans, and escaped slaves—lived in northern states. The South had only nine million people, many of them slaves who worked the land.

The North had other resources that gave it an advantage over the South. By the 1860s over 90 percent of the nation's factories were located in the North. Immigrants, eager to work for very low wages, filled the factories. Because of its steel mills and iron mines, the North could build ships and manufacture the weapons and tools necessary for warfare. Also, the textile mills in the Northeast had the waterpower to drive machines to produce cotton cloth for export, another important source of money. But the North got its cotton from the South, which depended on slaves for farming. The North did not believe in slavery and passed a law in 1808 forbidding the import of slaves.

A cotton plantation on the Mississippi by William Aiken Walker.

The South disagreed with the North on the issue of slavery, so it separated from the North and formed the Confederacy. The South believed it was a separate nation from the North and did not have to follow the North's law forbidding the import of slaves. It spent a great deal of money on its slave trade. Large farms, known as plantations, characterized the South.

Transportation, too, was growing more rapidly in the North than in the South. This was due, in part, to the North's abundance of industrial resources. The North had many miles of railroad, which made it easy for them to move food and supplies wherever they were needed for the war effort. The canal system connecting the North with the West was over 3,000 miles long. The South's transportation routes, however, did not connect many cities to each other.

Scene on the Erie Canal. Drawing, 1842.

The North had advantages over the South in agriculture also. With far more acres of farmland, the North grew a great variety of crops. The Midwest produced millions of bushels of wheat and corn and other grains. In addition, the labor-saving reaper and thresher permitted greater and faster harvests by fewer hands. The South, on the other hand, had fewer acres for farming. The South's main crop was cotton, but it also produced rice, sugar, and tobacco. It had to export these products in order to buy the ships and weapons needed for the war.

Resources such as factories, transportation, and the cultivation of many different crops gave the North a distinct advantage over the South as the threat of war approached.

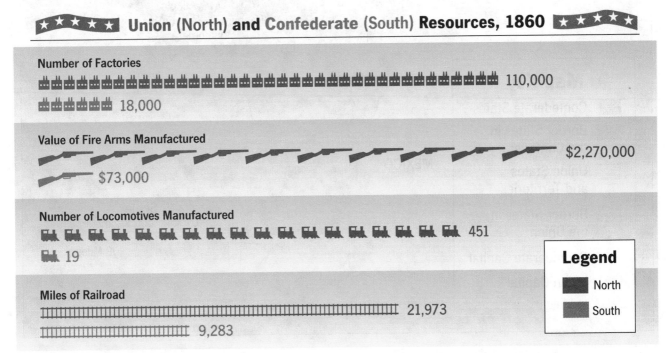

★ ★ ★ ★ ★ **Union (North) and Confederate (South) Resources, 1860** ★ ★ ★ ★ ★

Number of Factories
110,000
18,000

Value of Fire Arms Manufactured
$2,270,000
$73,000

Number of Locomotives Manufactured
451
19

Miles of Railroad
21,973
9,283

Legend
North
South

Historical maps show information about what happened in the past and where the events took place. For example, the conflict over slavery between the North and the South can be shown on a historical map. The Southern states were determined to maintain and even spread slavery. The North, however, wanted to end slavery. States were forced to choose sides over the issue of slavery.

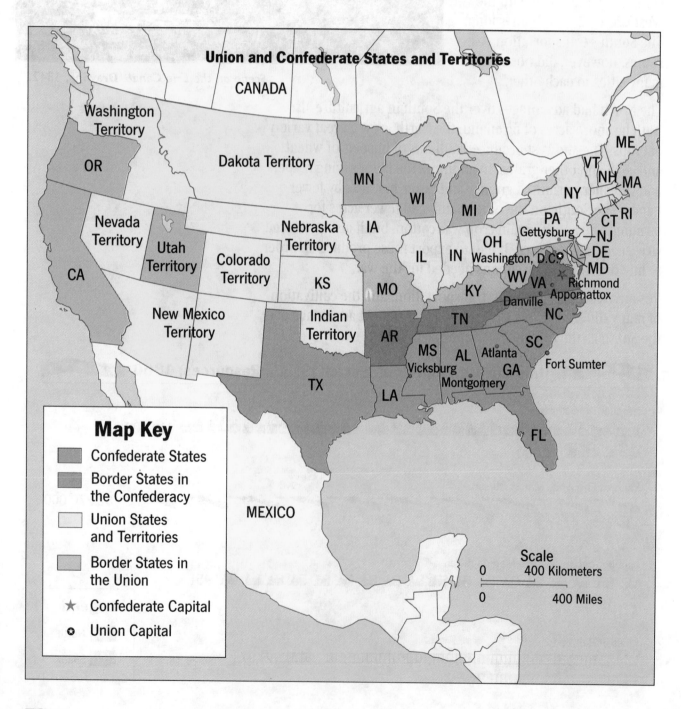

Union and Confederate States and Territories

CANADA

Washington Territory

OR

Dakota Territory

MN

WI

MI

ME

VT

NH

MA

NY

Nevada Territory

Utah Territory

Colorado Territory

Nebraska Territory

IA

IL

IN

OH

Washington, D.C.

PA

Gettysburg

NJ

DE

MD

CT

RI

CA

KS

MO

KY

WV

VA

Richmond

Appomattox

Danville

NC

New Mexico Territory

Indian Territory

AR

TN

SC

MS

AL

Atlanta

GA

Fort Sumter

Vicksburg

Montgomery

TX

LA

FL

MEXICO

Map Key

- ■ Confederate States
- ■ Border States in the Confederacy
- □ Union States and Territories
- ■ Border States in the Union
- ★ Confederate Capital
- ✪ Union Capital

Scale

0   400 Kilometers

0   400 Miles

A. Use the map to complete the following items about the North and the South.

1. What were the original Confederate states?

   TX, LA, MS, AL, GA, SC, FL

2. Which border states joined the Confederacy?

   AR, TN, VA, NC

3. What were the original Union states?

   OR, CA, UTAH, MN, WI, MI

4. Which border states sided with the Union?

   Washington, Dakota, Nevada, New Mexico, KS, IA

5. How many states fought for the Union?

   Six

6. How many states fought for the Confederacy?

   eleven

7. What city became the capital of the Union?

   PA

8. What city was the capital of the Confederacy?

   D.C

B. Think about the issue of slavery and the Civil War. Answer the following questions.

1. Did the North separate from the South or the South separate from the North?

   South seperated from the North

2. Why were the states divided into the North and the South?

   Because, they disagreed on slavery

3. Was a larger area of the land occupied by people who were for or against slavery?

   The people who were for slavery

Lesson 13

ACTIVITY   Explore and express your
opinion about the Civil War.

Civil War Resources

> ### The **BIG**
> ### Geographic Question
>
> **What situations did the northern and southern states face at the time of the Civil War?**

From the article you learned that at the time of the Civil War, the North had resources that gave it many advantages over the South. The map skills lesson showed you how the country was divided into the Union and the Confederacy over the issue of slavery. Now write a newspaper article about this important time in America's history.

A. List six major kinds of resources that both the North and the South needed in order to fight the Civil War.

1. _____

2. _____

3. _____

4. _____

5. _____

6. _____

B. Write a few sentences about one of the resources you listed above, comparing its availability and use in the North and the South and how it affected their readiness to fight a war.

_____

_____

_____

_____

C. Continue comparing the resources of the North and South by completing the following.

1. Name three industries that the North had.

_____

_____

_____

2. Why were railroads an important advantage for the North during the war?

_____

_____

_____

3. List two major crops that were produced in each part of the country.

North   South

_____   _____

_____   _____

4. Write a sentence describing one strength each side enjoyed at the start of the war.

North:

_____

_____

South:

_____

_____

D. Write an opinion editorial—a letter to a newspaper editor expressing your opinion about a topic—explaining who you think will win the war and why. Complete the following steps to help you organize your editorial.

1. Choose the viewpoint of either a northerner or a southerner at the beginning of the Civil War.

2. Include a headline that will capture your readers' attention.

3. Include in your first paragraph the important *who, what, when, where,* and *why* information.

4. Support your opinion with strong facts, reasons, and examples.

Lesson 14

Big City Bound!

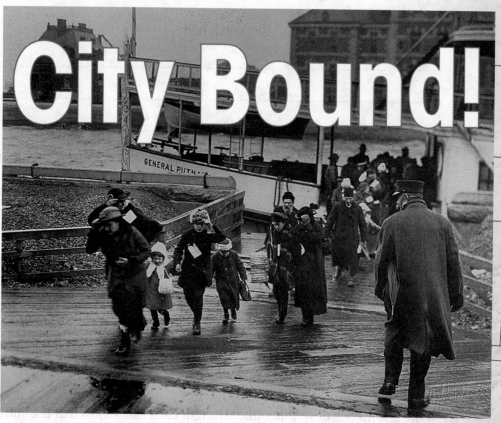

As you read about how cities grew after the Civil War, note how geography affected who moved to them and why.

Immigrants arrive at Ellis Island.

At the beginning of the Civil War in 1861, eight out of every ten people lived in rural areas. Most of them were farmers who were having trouble making a living. Many left their homes and went to the cities in hope of a better life. By the early 1900s, half of all Americans lived and worked in cities. This movement to and clustering in cities is called **urbanization.** It was a major event in the nation's history.

When the Civil War ended in 1865, millions of immigrants from all over Europe came to the United States. Some fled their homelands because of wars, religious persecution, overpopulation, and poverty. These reasons for leaving are "push factors." Others came simply to seek a better life. The belief that America offered a better life was a "pull factor."

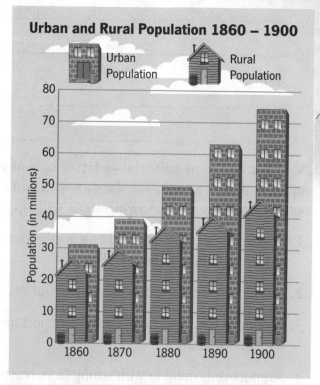

Between 1860 and 1900, America's urban population grew more rapidly than its rural population.

80

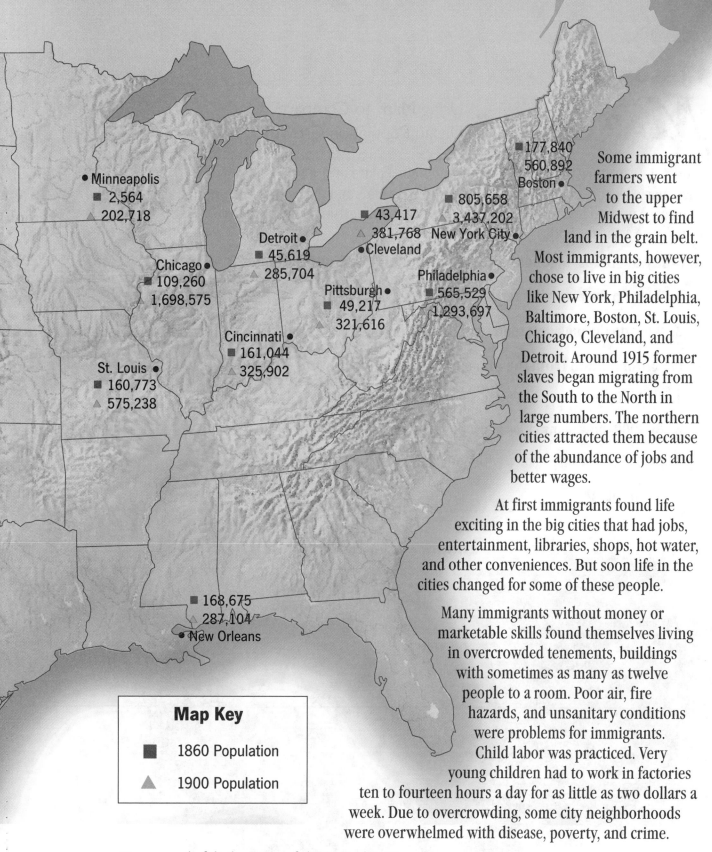

Some immigrant farmers went to the upper Midwest to find land in the grain belt. Most immigrants, however, chose to live in big cities like New York, Philadelphia, Baltimore, Boston, St. Louis, Chicago, Cleveland, and Detroit. Around 1915 former slaves began migrating from the South to the North in large numbers. The northern cities attracted them because of the abundance of jobs and better wages.

At first immigrants found life exciting in the big cities that had jobs, entertainment, libraries, shops, hot water, and other conveniences. But soon life in the cities changed for some of these people.

Many immigrants without money or marketable skills found themselves living in overcrowded tenements, buildings with sometimes as many as twelve people to a room. Poor air, fire hazards, and unsanitary conditions were problems for immigrants. Child labor was practiced. Very young children had to work in factories ten to fourteen hours a day for as little as two dollars a week. Due to overcrowding, some city neighborhoods were overwhelmed with disease, poverty, and crime.

Many people felt the cities of the United States were growing too rapidly. City officials had to figure out a way to solve these urgent problems. Soon politicians and community leaders began reforms that would eventually help improve the conditions of cities.

Map Key

■ 1860 Population

▲ 1900 Population

Minneapolis
■ 2,564
▲ 202,718

Detroit
■ 45,619
▲ 285,704

Chicago
■ 109,260
▲ 1,698,575

Cleveland
■ 43,417
▲ 381,768

Pittsburgh
■ 49,217
▲ 321,616

Cincinnati
■ 161,044
▲ 325,902

St. Louis
■ 160,773
▲ 575,238

Boston
■ 177,840
▲ 560,892

New York City
■ 805,658
▲ 3,437,202

Philadelphia
■ 565,529
▲ 1,293,697

New Orleans
■ 168,675
▲ 287,104

MAP SKILLS
Using Maps to Compare
Urban Population Growth

Two maps can show the changes in population in a place over a period of time. Each map's legend, or key, can help explain the symbols and the colors used on the map. Notice that when maps are used to make comparisons, the legend is usually the same for both.

A. Study the maps' legends and answer the following questions.

  1. What do the black dots represent? _____

  2. What do the pink dots represent? _____

B. Use information from the maps to answer the following questions.

  1. By 1870, how many cities had a population over 100,000? _____

  2. By 1900, how many cities had a population over 100,000? _____

  3. By 1870, how many cities with populations over 100,000 were located on the

  Great Lakes? _____

Growth of Cities by 1870

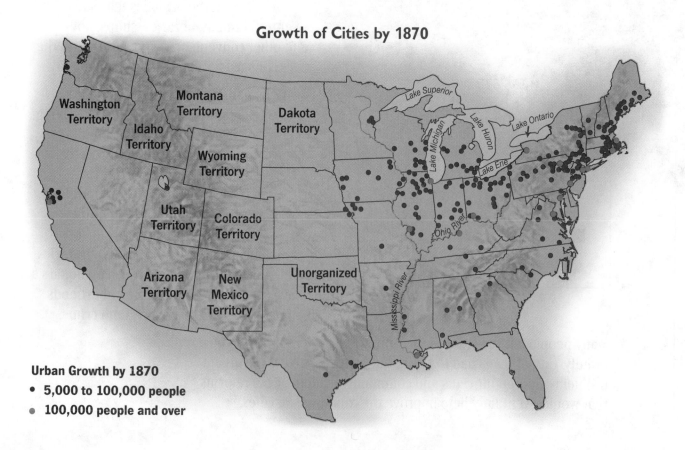

Urban Growth by 1870

● **5,000 to 100,000 people**

● **100,000 people and over**

4. By 1900, how many cities with populations over 100,000 were located on the Great Lakes?

_____

5. By 1870, how many cities with populations over 100,000 were located on the Mississippi River?

_____

6. By 1900, how many cities with populations over 100,000 were located on the Mississippi River?

_____

C. Use the maps and the above information to help you draw some conclusions about how the nation's cities grew between 1870 and 1900.

1. By 1870, where were most of the cities with populations over 100,000 located?

_____

2. By 1900, which area of the country had experienced the greatest growth in cities with populations over 100,000?

_____

Growth of Cities by 1900

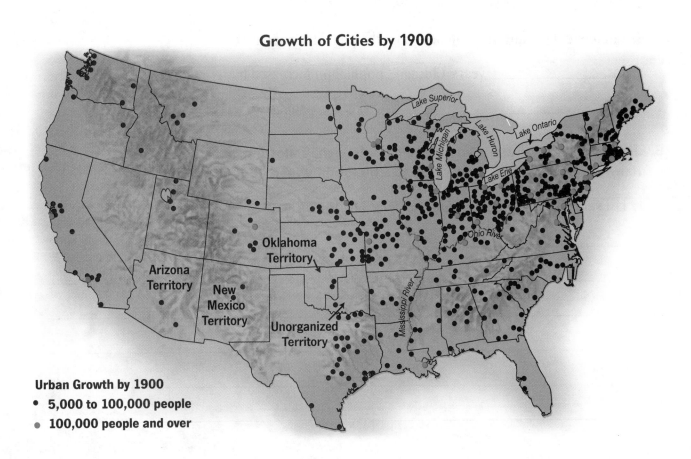

Urban Growth by 1900

- 5,000 to 100,000 people
- 100,000 people and over

Lesson 14

ACTIVITY
Find out about the effects of rapid growth on a city.

The Growth of a City

The **BIG**
Geographic Question

How did urban migration affect the city of Chicago?

You learned from the article that a great number of people moved from rural to urban areas in the United States after the Civil War. In the map skills lesson you looked at the growth in urban populations over time. Now find out how this enormous growth affected the city of Chicago.

A. Evaluate the rapid population growth of Chicago between 1860 and 1900. Use the article pages 80 and 81 to help answer the following.

1. What was Chicago's population by 1860? _____

2. What was Chicago's population by 1900? _____

3. Complete the following sentence about the population increase.

   In _____ years, Chicago's population increased by

   _____ people.

B. Use information in the Almanac to complete the following chart about the many ethnic groups that migrated to Chicago between 1860 and 1900.

1. List several of the ethnic groups that migrated to Chicago.

_____   _____

_____   _____

_____   _____

_____   _____

_____   _____

2. What were some of the reasons immigrants came to Chicago?

_____

_____

_____

3. Why did African Americans leave the South and go to Chicago?

_____

_____

_____

4. Describe the area and housing in which immigrants and African Americans moving from the South lived in Chicago.

_____

_____

_____

C. Describe how the land in Chicago was used during this time of sudden growth. Use the article, map, and Almanac to find out about transportation, parks, and modern conveniences in Chicago at the time.

_____

_____

_____

D. Charles Dickens wrote in his book _A Tale of Two Cities_, "It was the best of times, it was the worst of times . . . " Using the above information you have gathered about Chicago, describe how this quote applies to Chicago's urban growth.

_____

_____

_____

_____

_____

_____

Lesson 15

The POWER of WATERFALLS

As you read about glaciation and fall lines, think about how they affected the nation's geography.

Glaciers have sculpted much of North America. During the last Ice Age about 18,000 years ago, glaciers spread across North America. They covered the land as far south as the Ohio River. The glaciers scraped and scoured the land as they pushed soil and rock ahead of them, creating the rich layers of till that cover the Midwest today.

As Earth warmed and the glaciers melted, changes occurred in Earth's landscape. Deep gouges and gullies made by glaciers filled with water, creating lakes and streams. New rivers were formed and began to shape the landscape. In some places, such as the east coast of the United States, a series of almost parallel rivers spurt a line of waterfalls.

A waterfall is water flowing over a ledge.

A **fall line** develops as water moves from a higher elevation to a flat plain. When water erodes soft rock, it causes the water to flow over a ledge, forming a waterfall. In the United States, a major fall line stretches from New York to Alabama. Water from the Appalachian Mountains cuts into the soft rock of the Atlantic Coastal Plain. The Appalachian Mountains are the source of many rivers along this fall line. These rivers include the Potomac, the James, the Roanoke, the Chattahoochee, the Savannah, and the Susquehanna. Early colonists chose to settle in areas along the fall line because the rapids and falls made it impossible for their ships to travel farther inland. Cities such as Philadelphia, Pennsylvania; Richmond, Virginia; New York City, New York; Baltimore, Maryland; Columbia, South Carolina; Macon, Georgia; and Raleigh, North Carolina developed along the fall line.

Although the fall line may have prevented early colonists from seeking lands farther inland, it later benefited owners of lumber and textile mills. The rushing waters offered a source of energy for the mills. Industrialists built waterwheels over which the falls and rapids fell, providing the energy to run machinery.

The waterfalls and rapids along the fall line encouraged the growth of more mills and larger factories. During the mid-1800s many cities along the fall line from New York City to Baltimore became important centers of the Industrial Revolution. The Industrial Revolution introduced machines that brought changes in how goods were made and the volume that could be produced. Large volumes of goods could be produced in factories in a short amount of time. As a result, manufacturing became the fastest growing segment of the United States' economy during this period. The eastern fall line has greatly affected the growth of American cities and industries.

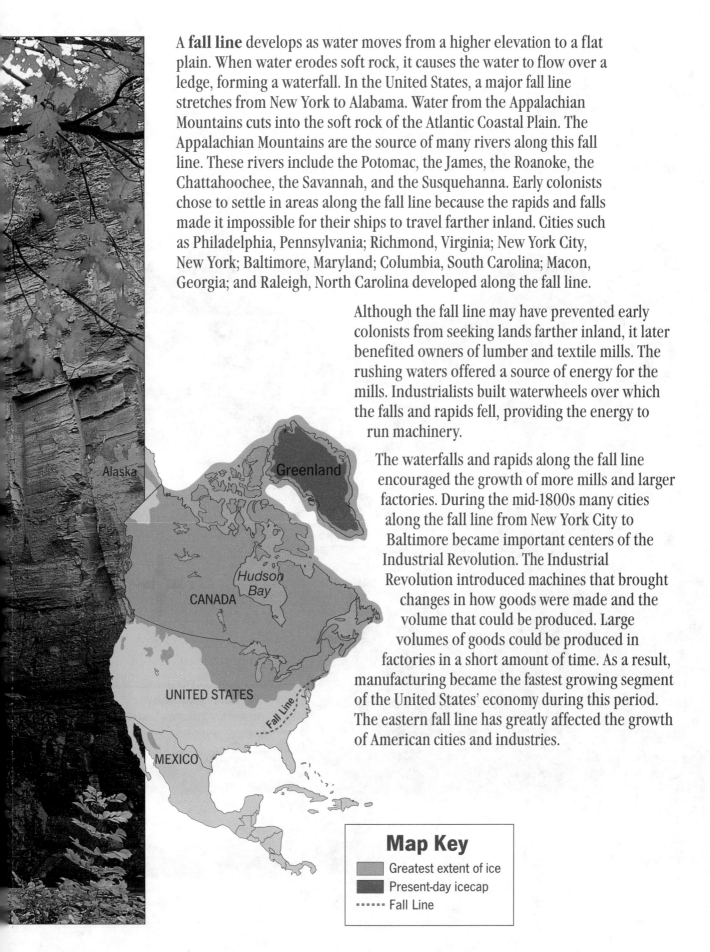

Alaska

Greenland

Hudson Bay

CANADA

UNITED STATES

Fall Line

MEXICO

Map Key

Greatest extent of ice

Present-day icecap

······ Fall Line

Lesson 15

MAP SKILLS Using a Relief Map

The word *relief* refers to the high points and low points of a land area.
A **relief map** shows different altitudes, or elevations, of mountains.
It also shows the relief of valleys.

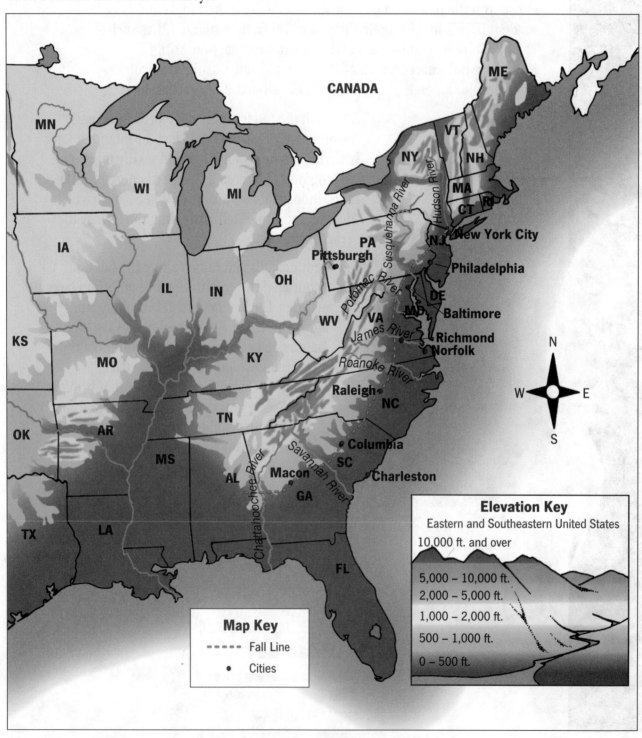

Elevation Key
Eastern and Southeastern United States
10,000 ft. and over
5,000 – 10,000 ft.
2,000 – 5,000 ft.
1,000 – 2,000 ft.
500 – 1,000 ft.
0 – 500 ft.

Map Key
- - - - - Fall Line
• Cities

A. List the major cities along the fall line.

_____   _____

_____   _____

_____   _____

B. Look at the map and write down the approximate elevations of the following cities.

  1. Raleigh, North Carolina _____

  2. Pittsburgh, Pennsylvania _____

  3. Baltimore, Maryland _____

  4. Columbia, South Carolina _____

C. Look at the fall line on the map. Answer the following questions.

  1. Explain why Raleigh, Baltimore, and Columbia are fall line cities.

_____

_____

_____

  2. Explain why Pittsburgh is not a fall line city.

_____

_____

_____

D. In the twentieth century, industry has developed in all regions of the United States. Explain why industry is no longer dependent on physical features such as the eastern fall line.

_____

_____

_____

_____

Lesson 15

CTIVITY

Find out how the Industrial Revolution was affected by the geography of the United States.

Geography and the Industrial Revolution

The **BIG** Geographic Question

What role did the eastern fall line and other geographical features play in America's Industrial Revolution?

From the article you learned how fall lines develop. In the map skills lesson you looked at fall lines and other geographical features on a relief map. Now learn more about what the eastern fall line had to do with the Industrial Revolution.

A. Answer the questions below to figure out how geography influenced the Industrial Revolution in the United States.

1. On the chart below list the cities that developed along the eastern United States fall line. Also, list the rivers associated with the eastern fall line.

Cities	Rivers

2. Why do you think New York is considered a fall line city?

_____

_____

_____

_____

3. If you were building a factory in the 1800s, would you build it along the fall line or in a rural area? Explain your answer.

_____

_____

_____

4. What natural resources (rivers, minerals, vegetation) would you look for in an area where you were building your factory? Explain why each resource is useful.

_____

_____

_____

_____

5. You have read about the effect of the eastern fall line on the industry of the region. Why did factories grow up along the fall line during the Industrial Revolution?

_____

_____

_____

_____

B. Make a plan for a 2- or 3-dimensional relief map that illustrates the eastern United States fall line.

1. Write your plan below.

_____

_____

_____

_____

_____

2. Decide what you will use to construct a raised relief map. You may use such materials as clay, aluminum foil, papier mâché, or construction paper. You may decide to use colors, shading, or contouring to show relief.

3. Construct your 2- or 3-dimensional relief map.

Industry "STEEL" Remains

As you read about the steel industry, think about how geography has affected its growth, change, and movement.

Pittsburgh was a military outpost during the French and Indian War (1754–1763). In the mid-1800s it became the birthplace of the United States' steel-making industry. It sat upon a large coal deposit and near three rivers. Railways and the waterways of the Great Lakes gave the city access to the best iron ore deposits. This combination made steel production one of its early industries.

In Pittsburgh, the Allegheny and Monongahela rivers meet to form the Ohio River. The Ohio River provided a way to transport steel to markets across the country. As a result, the steel industry in the United States grew rapidly in the late 1800s.

Railroads were helping to open the West to settlers, and steel was needed for trains and rails. Also, the factories that had started the Industrial Revolution in the United States needed steel for tools and machines. Later, the auto industry used enormous amounts of steel as did the United States military in the two world wars. By the early 1900s Pittsburgh had become the nation's largest steel producer, with 34 steel plants lining its riverbanks.

A worker wears protective clothing for safety in one of Pittsburgh's steel mills.

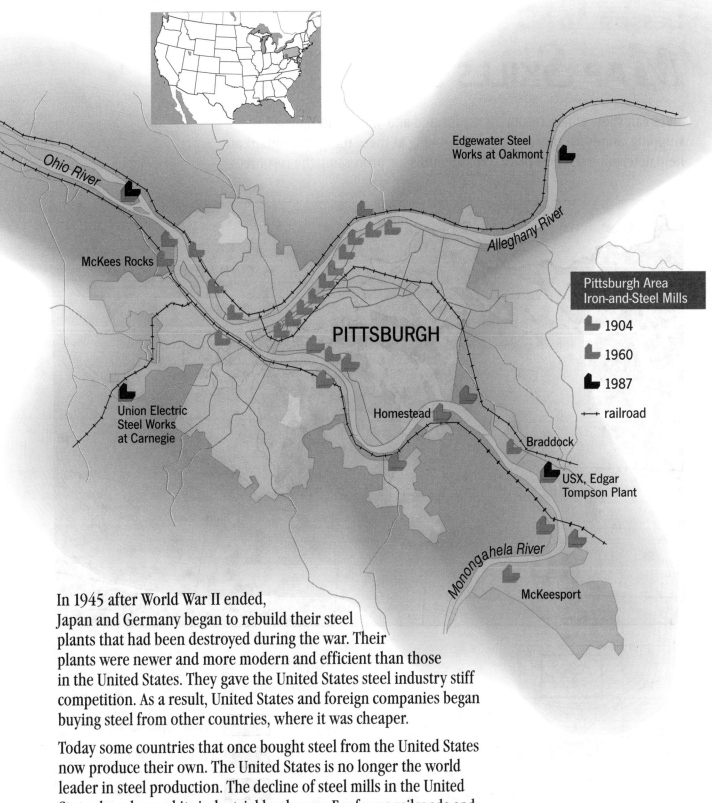

Pittsburgh Area Iron-and-Steel Mills

- 1904
- 1960
- 1987
- railroad

Ohio River

Edgewater Steel Works at Oakmont

Alleghany River

McKees Rocks

PITTSBURGH

Union Electric Steel Works at Carnegie

Homestead

Braddock

USX, Edgar Tompson Plant

Monongahela River

McKeesport

In 1945 after World War II ended, Japan and Germany began to rebuild their steel plants that had been destroyed during the war. Their plants were newer and more modern and efficient than those in the United States. They gave the United States steel industry stiff competition. As a result, United States and foreign companies began buying steel from other countries, where it was cheaper.

Today some countries that once bought steel from the United States now produce their own. The United States is no longer the world leader in steel production. The decline of steel mills in the United States has changed its industrial landscape. Far fewer railroads and factories are in use today. Land once occupied by mills now sits idle or is occupied by office buildings. Although Pittsburgh has new major industries, some steel is still produced there. The steel industry remains an important part of Pittsburgh's and our nation's history.

MAP SKILLS
Using a Map to Make
Decisions About Location

Maps can show where resources are located and water and railroad
shipping routes for those resources. Looking at these things on a map can
help determine the best place to locate mills for processing the resources.

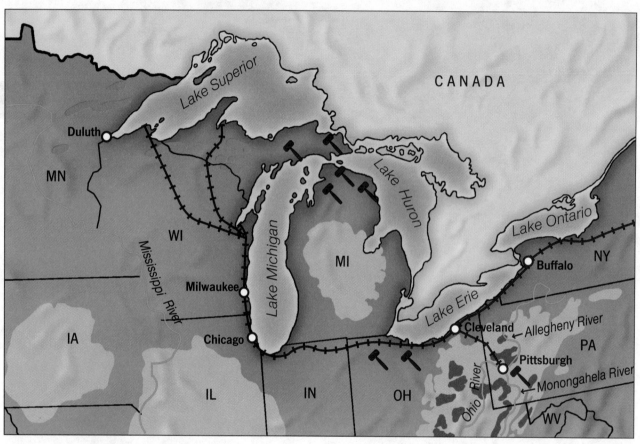

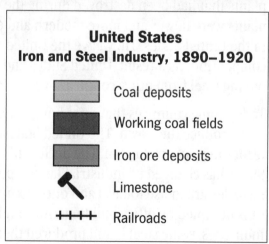

United States
Iron and Steel Industry, 1890–1920

Coal deposits

Working coal fields

Iron ore deposits

Limestone

Railroads

A. Look at the map and key and answer the following questions.

   1. What three mining resources does the map show?

     _____

   2. What two ways might you be able to ship these mining resources?

     _____

   3. If you were planning where to build a steel mill in Pittsburgh between 1890 and 1920, what would you need to consider?

     _____

B. Look at the location of the resources needed to produce steel and complete the following.

   1. Describe the location of the iron ore deposits. _____

     _____

   2. Describe the location of the working coal fields. _____

     _____

   3. Describe the location of limestone. _____

     _____

   4. What resource(s) would you need to transport and to what location? _____

     _____

     _____

   5. How would you transport the resource(s)? _____

     _____

C. Evaluate the information you have gathered above and decide where you would have built a steel mill in the early 1900s.

   1. Design and draw a steel mill logo, or symbol, on the map to show the location of your mill.

   2. Write an explanation of your decision below.

     _____

     _____

     _____

Lesson 16

Find out about the movement of the steel industry over time.

Movement and Change

The **BIG**
Geographic Question

Where was the United States steel industry originally located and why did it move over time?

From the article you learned how the steel industry began, grew, and declined. The map skills lesson showed you the location of the raw materials and the available shipping routes for Pittsburgh's iron and steel industry. Now find out what factors caused the location of the steel industry to change over time.

A. Use the article and Almanac to find out about various locations of the steel industry. Create a time line to show your information by completing the following.

1. From the 1860s to 1910

_____

2. From 1910 to the 1920s

_____

3. From the 1920s to the end of World War II in 1945

_____

4. From the end of World War II to the 1970s

_____

_____

5. From the 1970s to the present

_____

_____

_____

B. Use the article and Almanac to find information about the causes of change and movement in the steel industry in terms of the characteristics listed on the chart below. Complete the chart with the following questions in mind.

1. How did new technology increase the production of steel?

2. How did the way resources were used contribute to the movement of the steel industry from one location to another?

3. How did steel production in other countries affect the demand for U.S. steel?

Technology	Resources	Demand

C. Use the information you collected to create a visual display called "How the Steel Industry Has Changed over Time." Show the information from your chart in a time line, diagram, or poster of your own design. Use the space below to make a sketch of your visual display.

The Frontiers of TECHNOLOGY

As you read about advances in technology, think about some major breakthroughs that helped the United States expand.

The first mechanical cotton mill was built by **Samuel Slater** in **Rhode Island.**

In 1783 the United States was a medium-sized country hugging the coast of the Atlantic Ocean. Only 65 years later, the huge nation stretched from the Atlantic Ocean to the Pacific Ocean. Many new inventions helped Americans expand their nation.

Farm inventions helped pave the way westward. In New Haven, Connecticut, in1793 Eli Whitney invented a machine that could easily remove seeds from cotton. It was called the cotton gin. "Gin" is short for engine, and this engine made cotton planters very rich. Soon, a flood of southern farmers moved westward to start their own cotton plantations.

Many inventions were needed before settlers could farm the lands in the Great Plains region. This region lies west of the Mississippi River. Its prairie grasses had thick, deep roots that made it hard to plow the soil. A heavy-duty steel plow invented by John Deere in 1837 and a tough iron plow designed by James Oliver in 1869 solved this problem.

In addition, scientists developed a new farming method called "dry farming" to cultivate the drier regions of the Great Plains. Dry farming involved plowing deeper rows to bring underground water to the surface. After it rained, farmers loosened the soil to keep it moist longer.

On May 10, 1869, the first United States transcontinental railroad was completed.

The space shuttle has extended the frontier beyond Earth.

These farm inventions kept Americans moving west. Advances in transportation continued along with them. By 1811 steamboats on the Mississippi and Ohio rivers carried farm goods to market. With lots of money and hard work, states built canals to link the East and West. New York's Erie Canal opened in 1825 and served as a "gateway to the West." By 1869 the United States had its first **transcontinental railroad**–the Union Pacific Railroad–which met in Promontory, Utah. A transcontinental railroad is one that crosses a whole continent. In the twentieth century, Americans have continued to use technology to conquer different kinds of frontiers.

With the advances in technology made in recent centuries, people are no longer challenged by many of the physical difficulties of the past. The forbidding mountains, dense forests, and harsh climate are no longer barriers for people traveling west. Today travelers can fly from coast to coast in a matter of hours.

The newest frontier in technology is **cyberspace.** Cyberspace is a computer network that allows people all over the world to communicate with each other using telephone lines. This technology, along with many others, holds much promise for the high-tech explorers of the twenty-first century.

People no longer have to live and work near waterways and railways in order to travel or communicate from one place to another. Today's technology allows them to live and work almost anyplace they choose.

99

Lesson 17

MAP SKILLS Using Symbols to Create a Technology Map

Location plays an important role in the introduction of new technology. The obstacles associated with being in a certain location often inspire an inventor to create a new gadget. On the map below, identify the locations of some of the important technological inventions of the eighteenth, nineteenth, and twentieth centuries.

Technology Hot Spots

Legend

1. Using the article, the Almanac, and other reference materials, identify two advances in each field of technology and where they occurred.

Field of Technology	Technology Advancement	Location
Agriculture		
Transportation		
Communications		
Air and Space		
Computers		

2. Create a symbol for each of the above fields of technology. Your symbols can consist of a shape, a picture, or one or two letters.

3. Draw your symbols on the map legend. Be sure to label the technology field that each symbol represents.

4. For each invention you have listed, find its location on the map. Put a dot on the map where the city or town is located and label the dot with the city or town's name.

5. At the correct location, draw the symbol for each field of technology. This symbol should be big enough to see clearly but small enough so that your map does not look too crowded. You may choose to draw your symbols in the open space bordering the map and draw a line from the symbols to the city locations on the map.

6. Analyze your map of inventions and discuss with classmates why you think the inventions were made in those locations.

Lesson 17

ACTIVITY Create a time line showing advances in technology.

A Technology Time Line

The **BIG** Geographic Question

How did technology advance in the United States?

From the article you learned how technology helped expand the borders of the United States. The map skills lesson helped you identify the locations of different inventions. Now make a technology time line to show when these inventions occurred.

A. Make a list of important inventions that occurred between the 1700s and the present.

_____   _____

_____   _____

_____   _____

B. For the map skills lesson, you researched important inventions that helped the United States expand. You can use the same reference materials to find out the year when these inventions were created. Fill in the dates on the chart below. An example has been done for you.

Invention	Year Invented	Location	Significance
cotton gin	1793	New Haven, Connecticut	speeded up the removal of seeds from cotton so more could be produced and processed quicker

C. **Now use the information you wrote on the chart to complete a technology time line.**

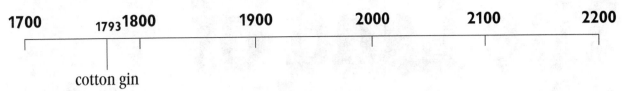

1700   1793 1800   1900   2000   2100   2200

cotton gin

D. **Evaluate the time line you completed and think about what new inventions will be created by the year 2100.**

1. Make a list of the inventions you think will be created in the future.

_____   _____

_____   _____

_____   _____

2. Use the space below to write a short paragraph about your predictions of inventions in the twenty-first century. Build the paragraph around the information that you collected on your charts and time line.

_____

_____

_____

3. Now add these invention predictions to your time line above.

4. Explain how these new technology advances will help the United States explore new frontiers.

_____

_____

_____

_____

_____

Lesson 18
The Land of Opportunity

As you read about the immigrants who came to the United States, think about the geographic and cultural features of the United States that attracted them.

During the Gold Rush of 1849, many Chinese immigrants came to California hoping to get rich in the California gold mines and then return to China.

"Fertile farmland! Lush forests! Plenty of open land for living! A fresh start!"

This is what the land of the United States was like in 1850. Millions of immigrants, people who settle in a country other than their homeland, from all over the world saw the United States as a land of wealth and freedom. One exception was the forced migration of millions of African blacks to work as slaves on plantations in the southern United States.

The perception of the United States as the land of opportunity was a "pull" factor for people of many different cultures. A pull factor attracts, or pulls people to voluntarily migrate from their native land to another country. The rich and varied resources of the United States attracted many immigrants. In the 1600s English immigrants settled in Virginia. They hoped to make their fortunes growing tobacco.

The inspiration for many other immigrant journeys to the United States often related to conditions in an immigrant's home country. These conditions are called "push" factors. In the 1850s the Irish potato crop was destroyed. Starvation pushed close to a million Irish farmers to the United States.

In Russia, Jewish villages were attacked in the late 1800s. This violence "pushed" many Russian Jews from their homeland, while the opportunity for a new life "pulled" them to the United States. In the late 1900s Mexican immigrants escaped from poverty and a violent civil war in their country. Immigrants from Norway, Sweden, and Denmark were driven from their homes by population pressure and a shortage of farmland. They staked out farms on the northern Great Plains, where the environment was very similar to their homelands.

Immigrants had to overcome many challenges. Upon arrival, many faced prejudice from native-born people of the United States as well as from many earlier immigrants. They had to adjust to a new environment, new language, and new customs far from home. Most overcame the challenges and chose to become United States citizens.

From its earliest days, the United States accepted immigrants from all over the world. There were periods in the United States' history, such as the 1930s, when immigration was restricted because of economic hardships in the United States. Today immigrants are facing some restrictions, but for the most part the United States is still a land of freedom, hope, and opportunity.

Dutch immigrants adapted to life on the Plains.

Lesson 18
MAP SKILLS Using a Map with a Graph to Show Migration

During the twentieth century, millions of immigrants made their way to the United States. The map below shows the oceans and continents they crossed. It is helpful to use graphs to track such large numbers of people. A graph is a diagram that shows numerical information in picture form.

Tides of Immigrants in the Twentieth Century

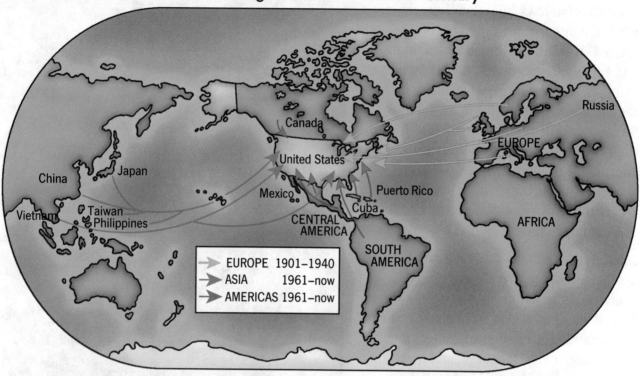

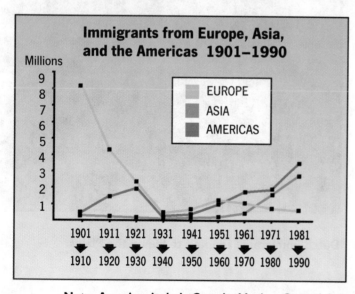

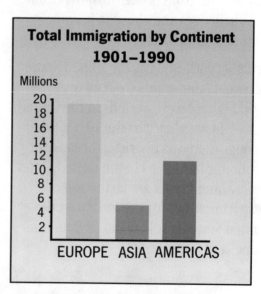

Note: Americas include Canada, Mexico, Central America, the Caribbean, and South America.

A. Look at the map of twentieth-century immigrants and the world map in the Almanac to complete the following.

1. List some of the countries in Europe from which immigrants came to the United States.

_____

_____

2. List some of the countries in Asia from which immigrants came to the United States.

_____

_____

3. List some of the countries in the Americas from which immigrants came to the United States.

_____

_____

B. Look at the two graphs and answer the following questions.

1. According to the bar graph on the right, what was the total number of immigrants that came to the United States from the following continents between 1901 and 1990?

   a. Europe _____ **b.** Asia _____ **c.** the Americas _____

2. According to the line graph on the left, what was the total number of immigrants that arrived in the United States from the following continents between 1911 and 1920?

   a. Europe _____ **b.** Asia _____ **c.** the Americas _____

C. The line graph shows how things change over a certain period of time. It also lets us compare the number of immigrants from Europe, Asia, and the Americas. Use the line graph to fill in the blanks below.

1. Between 1901 and 1910, _____ immigrants were the most numerous group.

2. The ten-year period between _____ and _____ was the first time that there were more Asian immigrants than European immigrants.

3. Between 1981 and 1990, immigrants from _____ were the most numerous group.

4. In the ten-year period between _____ and _____, immigration from Europe, Asia, and the Americas reached its lowest point.

Lesson 18

 **ACTIVITY** Find out about the people who have immigrated to the United States.

Life in the Land of Opportunity

The **BIG** Geographic Question

Is the United States the land of opportunity for immigrants?

From the article you learned about some of the push and pull factors that brought many immigrants to the United States. The map skills lesson used a map and graphs to show the numbers of immigrants who came to the United States in the early twentieth century. Now find information about immigrants who have come to the United States in the 1990s, and conduct a mock interview to show whether immigrants still see the United States as the land of opportunity.

A. Complete the following about immigrants who came to the United States in the late nineteenth and early twentieth centuries.

1. List some of the factors that pushed immigrants from other continents and countries to the United States.

a. _____   d. _____

b. _____   e. _____

c. _____   f. _____

2. What were some of the opportunities they might have expected to find in the

United States? _____

_____

_____

_____

_____

3. List some of the challenges that these immigrants faced once they arrived in the United States.

a. _____

b. _____

c. _____

d. _____

e. _____

B. Answer the following questions about people who are immigrating to the United States today.

1. Based on what you know about life today, do you think the United States is "a land of opportunity" for new immigrants coming here? Explain your answer.

_____

_____

_____

_____

2. Who are some of the groups of people who are immigrating to the United States today?

_____

_____

_____

_____

C. Interview an immigrant and compare his or her ideas about immigration with yours. Following are some questions you might ask.

1. Why did you leave your homeland?

2. How did you get to the United States? How long was your journey?

3. What opportunities did you expect to find in the United States? Did you find them?

4. Do you plan to stay or return to your home country?

5. Do you think the United States is the "land of opportunity"?

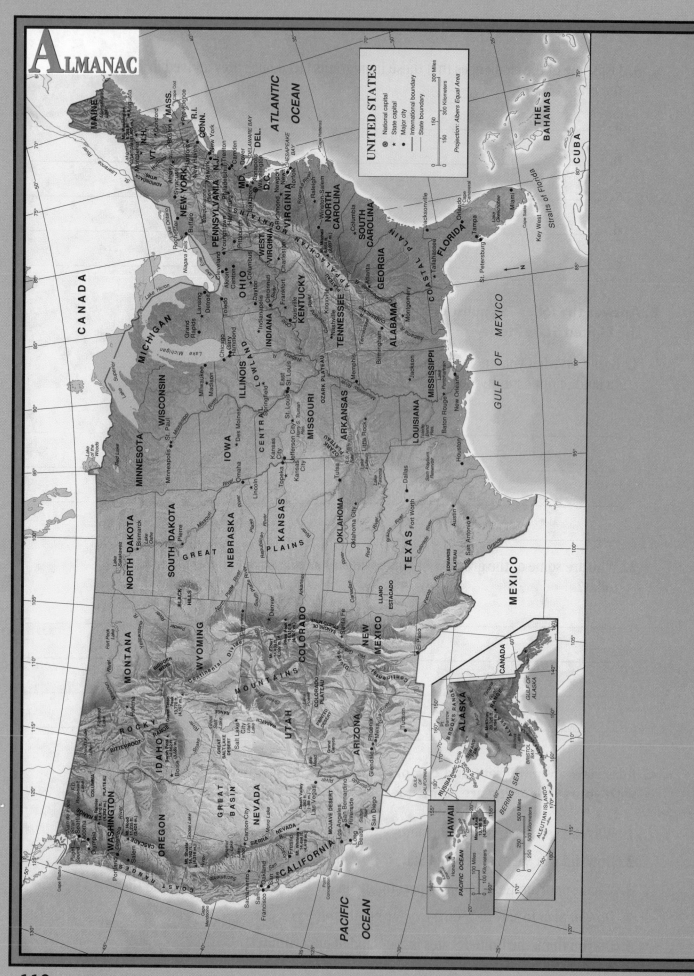

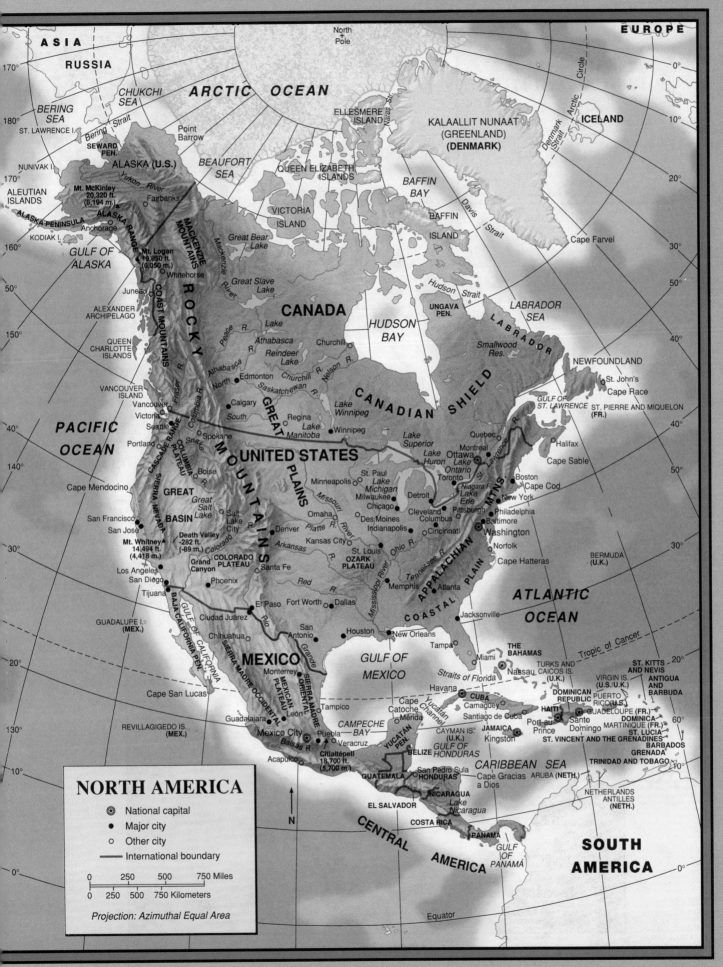

ASIA

170°

RUSSIA

180°

CHUKCHI SEA

BERING SEA

ST. LAWRENCE I.

170°

Bering Strait

Point Barrow

SEWARD PEN.

NUNIVAK I.

ALASKA (U.S.)

Mt. McKinley 20,320 ft. (6,194 m.)

Fairbanks

ARCTIC OCEAN

North Pole

ELLESMERE ISLAND

Nares Str.

EUROPE

10°

Arctic Circle

ICELAND

KALAALLIT NUNAAT (GREENLAND) (DENMARK)

Denmark Strait

20°

BEAUFORT SEA

QUEEN ELIZABETH ISLANDS

BAFFIN BAY

30°

ALEUTIAN ISLANDS

ALASKA PENINSULA

KODIAK I.

160°

Anchorage

ALASKA RANGE

GULF OF ALASKA

Mt. Logan 19,850 ft. (6,050 m.)

Whitehorse

Juneau

ALEXANDER ARCHIPELAGO

150°

VANCOUVER ISLAND

QUEEN CHARLOTTE ISLANDS

VICTORIA ISLAND

Great Bear Lake

MACKENZIE MOUNTAINS

Mackenzie River

Great Slave Lake

Peace R.

Lake Athabasca

Reindeer Lake

Athabasca R.

Churchill

North R.

Edmonton

Churchill R.

Nelson R.

Saskatchewan R.

CANADA

CANADIAN SHIELD

Smallwood Res.

BAFFIN ISLAND

Davis Strait

Cape Farvel

50°

Hudson Strait

UNGAVA PEN.

LABRADOR SEA

LABRADOR

40°

NEWFOUNDLAND

St. John's

Cape Race

PACIFIC OCEAN

40°

Vancouver

Victoria

Seattle

Portland

Spokane

Fraser R.

Columbia R.

South R.

Calgary

Regina

Lake Manitoba

Lake Winnipeg

Winnipeg

Lake Superior

HUDSON BAY

Quebec

Montreal

Ottawa

St. Lawrence R.

GULF OF ST. LAWRENCE

ST. PIERRE AND MIQUELON (FR.)

Halifax

Cape Sable

40°

Cape Mendocino

130°

San Francisco

San Jose

CASCADE RANGE

COLUMBIA PLATEAU

Snake R.

Boise

SIERRA NEVADA

GREAT BASIN

Great Salt Lake

Salt Lake City

Death Valley -282 ft. (-89 m.)

Mt. Whitney 14,494 ft. (4,418 m.)

ROCKY MOUNTAINS

GREAT PLAINS

UNITED STATES

Minneapolis

St. Paul

Milwaukee

Lake Michigan

Chicago

Omaha

Des Moines

Missouri R.

Detroit

Lake Huron

Lake Erie

Niagara Falls

Lake Ontario

Toronto

Cleveland

Pittsburgh

Columbus

Cincinnati

ADIRONDACK MTNS.

Boston

Cape Cod

New York

Philadelphia

Baltimore

Washington

50°

ATLANTIC OCEAN

30°

Los Angeles

San Diego

Tijuana

Denver

Colorado R.

Platte R.

Arkansas R.

Kansas City

St. Louis

Ohio R.

Tennessee R.

OZARK PLATEAU

Memphis

Atlanta

APPALACHIAN

Norfolk

Cape Hatteras

BERMUDA (U.K.)

30°

GUADALUPE I. (MEX.)

Santa Fe

COLORADO PLATEAU

Grand Canyon

Phoenix

Red R.

El Paso

Fort Worth

Dallas

Mississippi River

COASTAL PLAIN

Jacksonville

REVILLAGIGEDO IS. (MEX.)

Ciudad Juárez

Chihuahua

San Antonio

Houston

New Orleans

Tampa

Miami

THE BAHAMAS

Tropic of Cancer

20°

ST. KITTS AND NEVIS

MEXICO

Monterrey

MEXICAN PLATEAU

León

Tampico

GULF OF MEXICO

Straits of Florida

Nassau

Havana

TURKS AND CAICOS IS. (U.K.)

CUBA

VIRGIN IS. (U.S./U.K.)

PUERTO RICO (U.S.)

DOMINICAN REPUBLIC

GUADELOUPE (FR.)

ANTIGUA AND BARBUDA

DOMINICA

60°

Cape San Lucas

Guadalajara

SIERRA MADRE OCCIDENTAL

SIERRA MADRE ORIENTAL

Mexico City

Puebla

Veracruz

CAMPECHE BAY

Cape Catoche

Yucatán Channel

Mérida

Camagüey

Santiago de Cuba

CAYMAN IS. (U.K.)

JAMAICA

HAITI

Port-au-Prince

Santo Domingo

MARTINIQUE (FR.)

ST. LUCIA

ST. VINCENT AND THE GRENADINES

BARBADOS

BAJA CALIFORNIA PEN.

GULF OF CALIFORNIA

Balsas R.

Citlaltépetl 18,700 ft. (5,700 m.)

Acapulco

YUCATÁN PEN.

BELIZE

GULF OF HONDURAS

Kingston

CARIBBEAN SEA

ARUBA (NETH.)

GRENADA

TRINIDAD AND TOBAGO

10°

GUATEMALA

San Pedro Sula

HONDURAS

Cape Gracias a Dios

NETHERLANDS ANTILLES (NETH.)

EL SALVADOR

NICARAGUA

Lake Nicaragua

COSTA RICA

CENTRAL

AMERICA

PANAMA

GULF OF PANAMÁ

SOUTH AMERICA

0°

NORTH AMERICA

⊛ National capital

● Major city

○ Other city

— International boundary

| 0 | 250 | 500 | 750 Miles |

| 0 | 250 | 500 | 750 Kilometers |

Projection: Azimuthal Equal Area

N

Equator

0°

111

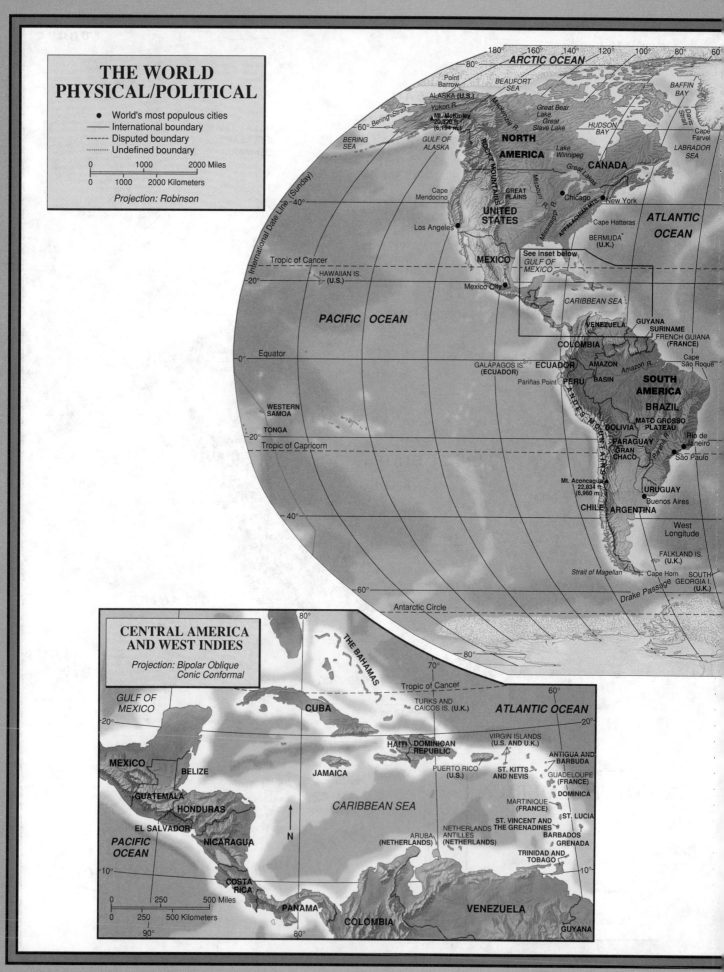

THE WORLD PHYSICAL/POLITICAL

- ● World's most populous cities
- —— International boundary
- ----- Disputed boundary
- ········· Undefined boundary

0   1000   2000 Miles
0   1000   2000 Kilometers

Projection: Robinson

ARCTIC OCEAN

Point Barrow
BEAUFORT SEA
BAFFIN BAY
ALASKA (U.S.)
Yukon R.
Great Bear Lake
Great Slave Lake
Davis Strait
Bering Strait
Mt. McKinley 20,320 ft. (6,194 m)
HUDSON BAY
Cape Farvel
BERING SEA
GULF OF ALASKA
ROCKY MOUNTAINS
Mackenzie R.
NORTH AMERICA
Lake Winnipeg
CANADA
Great Lakes
LABRADOR SEA
Cape Mendocino
GREAT PLAINS
Missouri R.
Mississippi R.
Chicago
New York
UNITED STATES
APPALACHIAN MTS.
ATLANTIC OCEAN
Cape Hatteras
International Date Line (Sunday)
Los Angeles
BERMUDA (U.K.)
Tropic of Cancer
MEXICO
See inset below
GULF OF MEXICO
HAWAIIAN IS. (U.S.)
Mexico City
CARIBBEAN SEA
PACIFIC OCEAN
VENEZUELA
GUYANA
SURINAME
FRENCH GUIANA (FRANCE)
COLOMBIA
Equator
GALÁPAGOS IS. (ECUADOR)
ECUADOR
AMAZON
Amazon R.
Cape São Roque
Pariñas Point
PERU
BASIN
SOUTH AMERICA
BRAZIL
WESTERN SAMOA
BOLIVIA
MATO GROSSO PLATEAU
TONGA
PARAGUAY
Rio de Janeiro
GRAN CHACO
Paraná R.
São Paulo
Tropic of Capricorn
Mt. Aconcagua 22,834 ft. (6,960 m)
URUGUAY
ANDES MOUNTAINS
CHILE
ARGENTINA
Buenos Aires
West Longitude
FALKLAND IS. (U.K.)
Strait of Magellan
Cape Horn
SOUTH GEORGIA I. (U.K.)
Antarctic Circle
Drake Passage

CENTRAL AMERICA AND WEST INDIES

Projection: Bipolar Oblique Conic Conformal

0   250   500 Miles
0   250   500 Kilometers

GULF OF MEXICO
THE BAHAMAS
Tropic of Cancer
TURKS AND CAICOS IS. (U.K.)
ATLANTIC OCEAN
CUBA
MEXICO
HAITI
DOMINICAN REPUBLIC
VIRGIN ISLANDS (U.S. AND U.K.)
ANTIGUA AND BARBUDA
BELIZE
JAMAICA
PUERTO RICO (U.S.)
ST. KITTS AND NEVIS
GUADELOUPE (FRANCE)
GUATEMALA
DOMINICA
HONDURAS
CARIBBEAN SEA
MARTINIQUE (FRANCE)
ST. LUCIA
EL SALVADOR
N
ST. VINCENT AND THE GRENADINES
BARBADOS
PACIFIC OCEAN
NICARAGUA
ARUBA (NETHERLANDS)
NETHERLANDS ANTILLES (NETHERLANDS)
GRENADA
TRINIDAD AND TOBAGO
COSTA RICA
PANAMA
VENEZUELA
COLOMBIA
GUYANA

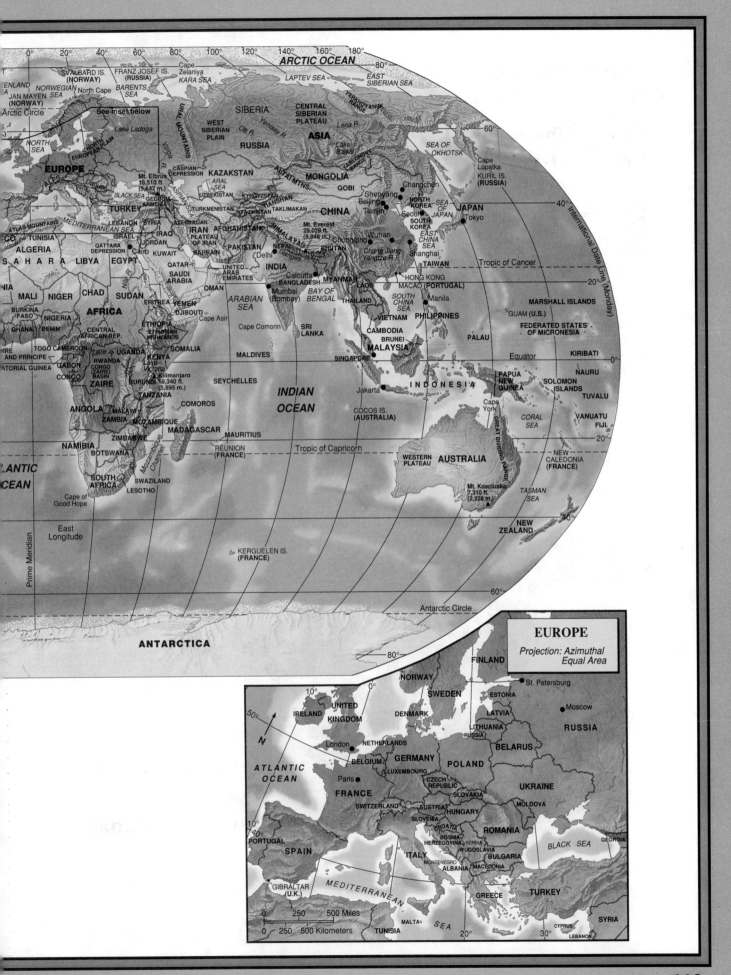

ARCTIC OCEAN

SVALBARD IS. (NORWAY)
FRANZ JOSEF IS. (RUSSIA)
Cape Zelaniya
KARA SEA
LAPTEV SEA
EAST SIBERIAN SEA
VERKHOYANSK RANGE

JAN MAYEN (NORWAY)
NORWEGIAN SEA
North Cape
BARENTS SEA
SIBERIA
CENTRAL SIBERIAN PLATEAU
Lena R.

Arctic Circle
See Inset below
NORTH SEA
Lake Ladoga
URAL MOUNTAINS
WEST SIBERIAN PLAIN
Ob R.
Yenisey R.
Lake Baikal
SEA OF OKHOTSK
Cape Lopatka
KURIL IS. (RUSSIA)

EUROPE
NORTH EUROPEAN PLAIN
ALPS
Volga R.
RUSSIA
ASIA
YABLONOVY RANGE

Mt. Elbrus 18,510 ft. (5,642 m.)
CASPIAN DEPRESSION
KAZAKSTAN
ARAL SEA
ALTAI MTNS.
MONGOLIA
Changchun
Shenyang
NORTH KOREA
SEA OF JAPAN
JAPAN

Danube R.
BLACK SEA
GEORGIA
ARMENIA
CASPIAN SEA
UZBEKISTAN
KYRGYZSTAN
TIANSHAN
GOBI
Beijing
Tianjin
Seoul
SOUTH KOREA
Tokyo

TURKEY
TURKMENISTAN
TAJIKISTAN
TAKLIMAKAN
CHINA
EAST CHINA SEA

LEBANON
SYRIA
AZERBAIJAN
IRAN
AFGHANISTAN
HIMALAYAS
Mt. Everest 29,028 ft. (8,848 m.)
Chongqing
Wuhan

ATLAS MOUNTAINS
MEDITERRANEAN SEA
ISRAEL
IRAQ
JORDAN
PLATEAU OF IRAN
PAKISTAN
NEPAL
Delhi
Ganges R.
BHUTAN
Chang Jiang (Yangtze R.)
Shanghai

CO
TUNISIA
ALGERIA
LIBYA
EGYPT
Cairo
KUWAIT
BAHRAIN
QATTARA DEPRESSION
Nile R.
QATAR
UNITED ARAB EMIRATES
INDIA
BANGLADESH
Calcutta
MYANMAR
LAOS
HONG KONG
MACAO (PORTUGAL)
TAIWAN
Tropic of Cancer

SAHARA
SAUDI ARABIA
OMAN
Mumbai (Bombay)
BAY OF BENGAL
THAILAND
SOUTH CHINA SEA
Manila

MALI
NIGER
CHAD
SUDAN
ERITREA
YEMEN
DJIBOUTI
ARABIAN SEA
VIETNAM
PHILIPPINES
GUAM (U.S.)
MARSHALL ISLANDS

BURKINA FASO
NIGERIA
AFRICA
CENTRAL AFRICAN REP.
ETHIOPIA
Cape Asir
Cape Comorin
SRI LANKA
CAMBODIA
BRUNEI
PALAU
FEDERATED STATES OF MICRONESIA

GHANA
BENIN
TOGO CAMEROON
ETHIOPIAN HIGHLANDS
SOMALIA
MALDIVES
MALAYSIA
SINGAPORE

IRE AND PRINCIPE
ATORIAL GUINEA
GABON
CONGO
ZAIRE
RWANDA
CONGO (ZAIRE) BASIN
UGANDA
KENYA
Lake Victoria
SEYCHELLES
INDONESIA
Equator
KIRIBATI
NAURU

BURUNDI
Kilimanjaro 19,340 ft. (5,895 m.)
TANZANIA
INDIAN OCEAN
Jakarta
PAPUA NEW GUINEA
SOLOMON ISLANDS
TUVALU

ANGOLA
MALAWI
ZAMBIA
COMOROS
MADAGASCAR
MAURITIUS
Cape York
CORAL SEA
VANUATU
FIJI

NAMIBIA
ZIMBABWE
BOTSWANA
Mozambique Channel
RÉUNION (FRANCE)
Tropic of Capricorn
WESTERN PLATEAU
AUSTRALIA
GREAT DIVIDING RANGE
NEW CALEDONIA (FRANCE)

ATLANTIC OCEAN
SOUTH AFRICA
SWAZILAND
LESOTHO
Cape of Good Hope
Mt. Kosciusko 7,310 ft. (2,228 m.)
TASMAN SEA

East Longitude
NEW ZEALAND

Prime Meridian
KERGUELEN IS. (FRANCE)

Antarctic Circle

ANTARCTICA

International Date Line (Monday)

EUROPE

Projection: Azimuthal Equal Area

FINLAND
NORWAY
SWEDEN
St. Petersburg
ESTONIA
Moscow

IRELAND
UNITED KINGDOM
DENMARK
LATVIA
LITHUANIA
RUSSIA
RUSSIA

London
NETHERLANDS
BELGIUM
GERMANY
POLAND
BELARUS

ATLANTIC OCEAN
Paris
LUXEMBOURG
CZECH REPUBLIC
SLOVAKIA
UKRAINE

FRANCE
SWITZERLAND
AUSTRIA
HUNGARY
SLOVENIA
MOLDOVA

N

PORTUGAL
SPAIN
CROATIA
BOSNIA-HERZEGOVINA
SERBIA
YUGOSLAVIA
ROMANIA

ITALY
MONTENEGRO
ALBANIA
MACEDONIA
BULGARIA
BLACK SEA
GEORGIA

GIBRALTAR (U.K.)
MEDITERRANEAN
GREECE
TURKEY
SYRIA

0 250 500 Miles
0 250 500 Kilometers
MALTA
TUNISIA
SEA
CYPRUS
LEBANON

113

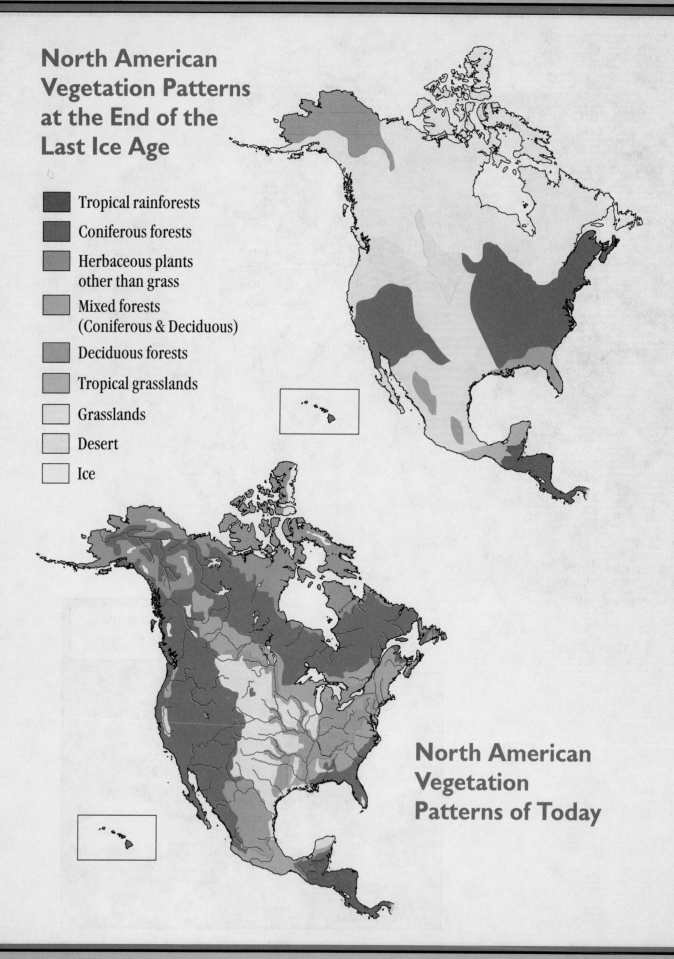

North American Vegetation Patterns at the End of the Last Ice Age

- Tropical rainforests
- Coniferous forests
- Herbaceous plants other than grass
- Mixed forests (Coniferous & Deciduous)
- Deciduous forests
- Tropical grasslands
- Grasslands
- Desert
- Ice

North American Vegetation Patterns of Today

California

Colorado

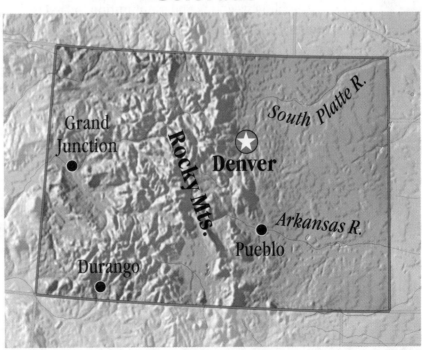

Battles of the Revolutionary War

Yorktown	The battle of Yorktown took place in southeastern Virginia on a peninsula that extends into Chesapeake Bay. The peninsula is also surrounded by the James and York rivers. British General Charles Cornwallis took his men to Yorktown to build fortifications. It was during this time that the French and the American armies realized that the location Cornwallis had chosen was an inviting target because it was open to attack from the water. A small American force positioned itself on land to prevent Cornwallis' escape from the peninsula. Meanwhile, American and French naval forces positioned themselves on the James River and in Chesapeake Bay. Cornwallis' location on a peninsula allowed French and American regiments to successfully surround him and his men. A last attempt by Cornwallis to escape by boat was spoiled when a storm arose and swamped the small boats. The General and his men eventually surrendered to the American and French troops.
Lexington and Concord	The battles of Lexington and Concord took place in eastern Massachusetts, just west of Boston. The surrounding land area is defined as the coastal lowlands, and is covered with rounded hills, swamps, small lakes, and shallow rivers. British General Thomas Gage knew from his informers that American troops were assembling weapons and ammunition in Lexington and Concord. He was also told that they had carried off a hundred pounds of gunpowder that belonged to the crown. General Gage received orders from the British government to take military action against the Americans. However, the Americans learned about the orders before General Gage did. When Gage and his men marched into Lexington, the American minutemen were prepared. No one knows who fired the first shot, but eight minutemen were killed, and ten more were wounded. One British soldier had been hurt. The British continued on to Concord, where they searched for hidden weapons and ammunition. Some of Gage's men met minutemen just outside of Concord. After a brief clash the British turned back toward Boston. Along the way, Americans fired at them from behind trees and stone fences. About 250 British were killed and wounded, while American losses came to about 90.

Battles of the Revolutionary War

Trenton

The battle of Trenton took place in western New Jersey on the Atlantic coastal plain. The coastal plain is a gently rolling lowland. Its eastern regions are covered with pine forests and salt marshes. The Delaware River flows through the western portions of the coastal plain.

On December 25, 1776 General George Washington led his army, exhausted and starving, to battle against the British in Trenton, New Jersey. Washington's troops were on the Pennsylvania side of the Delaware River, and the British had taken over Trenton. Washington's plan involved three lines, and used the surrounding physical features. The first line had men crossing the Delaware River downstream and engaging the British as a diversion. The second line involved men crossing the river to hold the bridge and prevent a British escape from Trenton. Washington was to lead the third line of men across the Delaware River nine miles upstream and advance on Trenton from the north.

Another physical feature that influenced the battle was the weather conditions at Trenton. General Washington was unaware that the other lines had found the Delaware River too fast and cold to cross. However, Washington and his line did cross the river. It took the men most of the night through driving sleet and cold temperatures.

After Washington's troops crossed the river, they attacked from the east and from the north. The British were caught unprepared to fight and began to flee. However, they were unable to escape because some of Washington's men held the bridge over the river. As a result, many British were captured and some were killed.

Source: *World Book Encyclopedia*, 1991; *Rand McNally Atlas of the American Revolution*, Kenneth Nebenzahl, editor, 1974.

Chicago Highlights

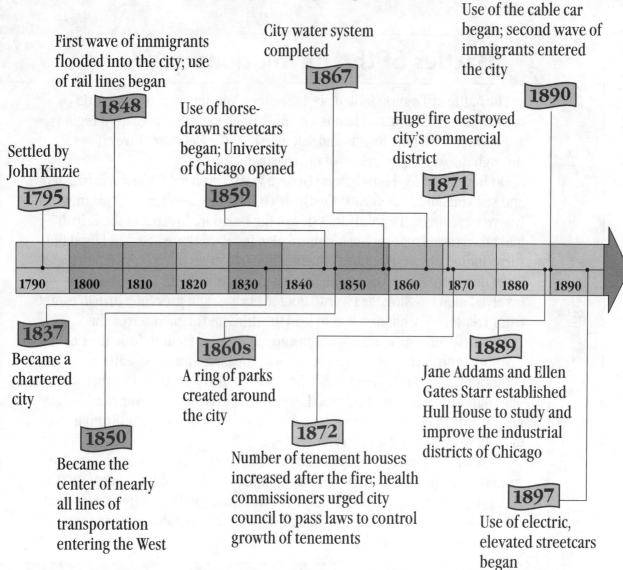

First wave of immigrants flooded into the city; use of rail lines began
1848

City water system completed
1867

Use of the cable car began; second wave of immigrants entered the city
1890

Use of horse-drawn streetcars began; University of Chicago opened
1859

Huge fire destroyed city's commercial district
1871

Settled by John Kinzie
1795

1790 1800 1810 1820 1830 1840 1850 1860 1870 1880 1890

1837
Became a chartered city

1850
Became the center of nearly all lines of transportation entering the West

1860s
A ring of parks created around the city

1872
Number of tenement houses increased after the fire; health commissioners urged city council to pass laws to control growth of tenements

1889
Jane Addams and Ellen Gates Starr established Hull House to study and improve the industrial districts of Chicago

1897
Use of electric, elevated streetcars began

Ethnic Groups	Population		Geographic Features
Czechoslovakians	1840	4,470	• located on the shores of Lake Michigan
Germans	1850	29,963	• the Chicago River runs through the city
Irish	1860	112,172	• fertile land can be found to the west of the city
Italians	1880	500,000	• many parks can be found in and around the city
Norwegians	1990	2,783,726	• a rapid transit system loops around the city
Poles	1995*	2,732,300	
Russian Jews	* estimated population as of January 1, 1995		
Swedes			

Sources: *National Geographic Historical Atlas of the United States.* 1993; *1996 Information Please Almanac*, Otto Johnson, editor.

Features of Possible North American Settlement Sites in the 1700s

Site	Climate	Land and Vegetation	Waterways	Natural Resources
San Antonio, Texas	• 111 days per year are over 90°F; average summer temperature is 84°F; average winter temperature is 52°F; average annual precipitation is 28 inches	• desert dotted with rock; deep canyons; southern Rocky Mountain ranges • brush country with pear cactus, buffalo grass, and bunchgrass	• San Antonio River	• fertile coastal plains soil; fruit and vegetables
Philadelphia, Pennsylvania	• average summer temperature is 71°F; average winter temperature is 43° F; average annual precipitation is 41.5 inches	• fertile pastures; coastal plains; hills of the Piedmont Plateau; Appalachian Mountains • forests of oak, beech, maple, walnut, pine, and ash trees	• Delaware and Schuylkill rivers meet; bays and inlets	• fertile coastal plains soil; rivers; timber from forests
New Orleans, Louisiana	• average summer temperature is 77°F, and winter average is 60°F; average annual precipitation is 60 inches	• coastal plains; southern Appalachian Mountain ranges • bayou, marshes, and swamps; Spanish moss; beech and walnut trees	• Mississippi River; Gulf of Mexico	• marshes and bayous filled with wild life such as ducks, geese, alligators, and beavers; forests; fertile coastal plains soil

Regional Native American Tribes

Great Plains Native Americans	Iowa, Kansa, Missouri, Wichita, Comanche, Omaha, Crow, Cheyenne, Sioux, Lakota

Spotlight on: Cheyenne Tribe	
Region	North American Plains near the Platte and Arkansas rivers; in the Black Hills of South Dakota; tall grasslands and rivers
Climate	warm, dry summers; cold winters with blowing snow
Shelter	tepees were made of long poles and buffalo skins which could be easily taken apart and moved
Resources	buffalo used for fuel, food, and clothing; fertile river valley allowed for some farming

Northeast Native Americans (Eastern Woodlands)	Mohawk, Cayuga, Seneca, Oneida, Onondaga, Iroquois, Delaware

Spotlight on: Seneca Tribe	
Region	western New York and eastern Ohio
Climate	warm summers; cold, snowy winters
Shelter	longhouses built of wood and bark
Resources	Fertile soil and an abundance of fresh water from the Great Lakes allowed them to cultivate corn and other vegetables; hunted a variety of wild game

Northwest Native Americans	Makah, Haida, Okanagoh, Quinault, Nootka, Chinook, Spokane, Kalapuya, Kalispel, Shuswap

Spotlight on: Makah Tribe	
Region	Northwest Pacific coast of North America; Vancouver Island at northwest tip of Washington state
Climate	mild temperatures and heavy rainfall
Shelter	wooden plank houses
Resources	forests for shelter; salmon, whale, and caribou used for food

Southeast Native Americans	Cherokee, Chickasaw, Choctaw, Creek, Seminole, Powhatan, Natchez, Timicua, Sauk, Caddo

Spotlight on: Cherokee Tribe	
Region	Southern Appalachian region; North Carolina and Tennessee
Climate	mild seasons: warm summers and wet winters
Shelter	wigwams: huts with arched frameworks of poles covered with bark or animal hides
Resources	rich soil for farming crops of corn, beans, and squash, along with hunting wild game such as deer and birds

Southwest Native Americans	Navajo, Apache, Hopi, Pima, Pueblo, Papago, Cochimi

Spotlight on: Navajo Tribe	
Region	Four Corners—where Utah, Colorado, Arizona, and New Mexico meet; desert terrain with many cliffs, canyons, and rock formations
Climate	hot and dry with little rain
Shelter	hogans: houses built of logs and mud
Resources	desert soil good for growing crops such as beans, cotton, and corn; also used to make clay for pottery

Ethnic Composition of New York City

(approximate % of total population)

Ethnic Group	1990
White	52.26
African American	28.71
Native American	.38
Asian and Pacific Islander	7.0
Hispanic	24.36
Other	11.65

Source: *World Encyclopedia of Cities,* George Thomas Kurian, 1994

Immigrants to the United States in the 1900s

Recent immigrants to the U.S. come from many regions of the world

Asia	
China	South Korea
Hong Kong	India
Thailand	Pakistan
Vietnam	

Middle East	
Israel	Syria
Jordan	Saudi Arabia

Latin America
Mexico
Cuba
Haiti

Europe
the former Soviet Union
Bosnia

Population Terms and Definitions

Population	the total number of people that live in a particular place or area • Example: The population of Idaho as of 1990 was 1,006,749. The population of Rhode Island as of 1990 was 1,003,464.
Population Distribution	the way in which people are scattered or spread out over an area
Population Density	the number of people per unit (for instance, per square mile) that live in an area or place • Example: The population density for Idaho as of 1990 was 12.2 people per square mile. The population density for Rhode Island as of 1990 was 951.1 people per square mile.
High Population Density	a large number of people living in an area; a densely populated area • Example: Rhode Island has a high population density.
Low Population Density	a small number of people living in an area; a sparsely populated area • Example: Idaho has a low population density.

Idaho

Rhode Island

Push/Pull Factors for Human Movement

Push Factors	Characteristics of Europe that Led Colonists to Leave	
1. shortage of land	4. political oppression	
2. religious intolerance	5. overcrowded cities	
3. lack of economic opportunity	6. crowded prisons	
Pull Factors	**Characteristics of North America that Enticed Europeans to Come**	
1. land and resources	4. land suitable for farming	
2. freedom of religion	5. freedom from political oppression	
3. economic opportunities	6. opportunities for a new start	

Source: *1996 Information Please Almanac*, Otto Johnson, editor; *Human Geography: Landscapes of Human Activities*, J.D. Fellman, 1992.

Exchanges Between the Old World and the New World

From the Old World to the New					
Animals	**Food**				**Diseases**
horses	wheat	soybeans	lettuce	citrus fruits	smallpox
cattle	rice	sugarcane	peaches	rye	measles
pigs	barley	onions	pears	olives	diptheria
sheep	oats	bananas	watermelons	chickpeas	
chickens					
honeybees					

From the New World to the Old					
Animals	**Food**				**Flowers/Plants**
turkeys	corn	avocados	peppers	pineapples	sunflowers
	tomatoes	cashews	chocolate	wild rice	marigolds
	beans	blueberries	tobacco	sweet potatoes	petunias
	vanilla	squash	cassava		
	pumpkin	potatoes	peanuts		
Settlers in North America also found quinine, a cure for malaria					

124

Trails Across the Western Frontier

	Physical Features	Native Peoples	Food	Difficulties Along the Way
Oregon Trail	• rugged terrain of the Black Hills; Rocky Mountains; dry, sandy areas with rough, rocky ground; swift rivers and streams; trees and bunch grasses	• Sioux Native Americans • traded moccasins and beads for bread • sold salmon to pioneers • used buffalo dung for fuel	• wild game, including buffalo • mountain goats and lions • whales and otters	• scarlet fever, mumps, cholera, mosquito fever; rattlesnakes, disagreements with Native Americans
National Road	• Ohio, Scioto, Miami, and Wabash rivers; Appalachian Mountains; trees, wild onions, and mushrooms	• Shawnee and Miami Native Americans	• wild game, including deer • rabbits and wild turkeys	• steep slopes, narrow roads, and sharp turns of the Appalachian Mountains
Mormon Trail	• plains of Iowa and Nebraska; Platte River; Rocky Mountains; trees, grasses, wild flowers, and berries	• Arapaho, Omaha, and Iowa Native Americans	• wild game, including buffalo, deer, and elk	• scarlet fever, cholera, rattlesnakes, disagreements with Native Americans
California Trail	• Humboldt River; Nevada desert; Sierra Nevada Mountains in California; tumble-weeds, cactus, prairie grasses, and trees		• mountain goats, coyotes, and other wild game	• cholera and dysentery; snowstorms
Santa Fe Trail	• tall grasses over six feet high; Jornada del Muerto desert; Rocky Mountains; trees	• Comanche Native Americans • used buffalo dung for fuel	• buffalo, deer, antelope, and other wild game	• got lost in the tall grasses; harsh desert conditions

Sources: *The Story of America's Roads*, Ray Spangenburg and D. M. Moser 1992; *From Trail to Turnpikes*, Tim McNeese, 1993; *World Book Encyclopedia*, 1991.

World Explorers

Marco Polo	Venice, Italy	Shangdu, China	7,500 miles	1271–1274	boat and camel	3 years
Christopher Columbus	Spain	West Indies	4,500 miles	1492	caravel	2 months, 9 days
Giovani Caboto (John Cabot)	Britain	North American mainland	500 miles	1497	small caravel	33 days
Vasco da Gama	Lisbon, Portugal	India	12,000 miles	1497–1498	caravel	10 months
Amerigo Vespucci	Portugal	South America	6,000 miles of coast	1501	caravel	3 months
Vasco Nunez de Balboa	Hispaniola	Pacific Ocean	1,000 miles	1513	two-masted sailing ship	32 days
Ponce de Leon	Spain	Florida	1,100 miles	1513	caravel	24 days
Hernando Cortes	Cuba	Mexico	1,600 miles	1519	caravel	35 days
Ferdinand Magellan	Spain	sailed around South America and back to Spain	50,000 miles	1519–1521	caravel	3 years
Jacques Cartier	France	St. Lawrence River	5,000 mles	1535	three caravels	2 months, 23 days
Francisco Coronado	Spain	Southwest U.S.	3,300 miles	1540–1542	foot; horseback	2 years
Samuel de Champlain	France	Maine coast	5,500 miles	1604–1607	caravel	3 years

E						
Robert Fulton	United States	Albany, NY	150 miles	1807	steamboat	30 hours
Nathaniel Palmer & Fabian Gottlieb Von Bellingshausen	United States & Russia	Antarctica	700 miles from the tip of South America	1820–1821	sloop (small vessel with triangular sails)	2 years
Robert E. Peary	United States	North Pole	450 miles	1909	dog sled	31 days
Calbraith Rodgers	United States	New York to California	3,000 miles	1911	airplane	84 days
Charles Lindbergh	United States	New York to Paris	3,610 miles	1927	airplane	33 1/2 hours
Amelia Earhart	United States	first woman to fly solo across Atlantic Ocean	2,026 miles	1932	airplane	15 hours
Yuri Gagarin	Soviet Union	First human in space	1 orbit of Earth, 203 miles above	1961	Vostok rocket	89 minutes
Neil Armstrong & Edward Aldrin	United States	First humans to walk on the moon	215,000 miles from Earth	1969	spacecraft and lunar module	4 days
Clay Lacy	United States	Flew around the world	26,345 miles	1988	airplane	37 hours
Shannon Lucid	United States	first woman to spend extended time in space	240 miles above Earth	1996	space shuttle and space station	6 months
ordinary citizens	the world	flight from New York, NY to London	3,469 miles	1996	supersonic transport	3 hours

Sources: *World Explorers and Discoverers*, Richard E. Bohlander, editor, 1992.; *Explorers and Discoverers of the World*, Daniel B. Baker, editor, 1993.

Explorers to Western Regions of the United States

Lieutenant Zebulon Pike 1805–1807

In August 1805 Lieutenant Zebulon Pike led an expedition of twenty soldiers from St. Louis, Missouri, up the Mississippi River. The purpose of his mission was to find the source of the great river. Along the way, Pike purchased 100,000 acres of land in Minnesota from the Sioux Native American tribe, an area which later became known as Minneapolis. Pike and his men reached Red Cedar and Leech lakes in February 1806. Pike wrongly believed that these lakes were the source of the Mississippi. The true source was Lake Itasca. Pike returned to St. Louis in April 1806.

In July 1806 Pike set out on a second expedition to search for the sources of the Arkansas and Red rivers. Along with 23 soldiers and 51 Osage Native Americans, Pike traveled up the Missouri River to the Osage River. Using horses, they then traveled through Osage lands to the Republican River in western Nebraska. In November 1806 Pike's expedition saw the front range of the Rocky Mountains, near what is now Pueblo, Colorado. Pike attempted to climb the highest eastern-most peak. However, lack of warm clothing and supplies forced him to turn back. This 14,110 foot mountain is named Pike's Peak for Lieutenant Pike.

The expedition returned to the Arkansas River, tracing it to its source in the Royal Gorge in south central Colorado. From here, Pike and his men attempted to reach the Red River. They crossed through the southern portion of the Rocky Mountains. In the spring of 1807 Pike's party was arrested for illegally crossing into Spanish territory. After being held in Mexico for several months, they returned to the United States on June 30, 1807.

John Wesley Powell 1869

On May 24, 1869 John Wesley Powell and a party of eleven men began an expedition of the Colorado River. They began their boat journey near the town of Green River, Wyoming. Reaching the Colorado River, the expedition soon entered the Grand Canyon. Three of Powell's men decided not to continue, so Powell and the remaining eight men proceeded down the rapids and through the mile-high walls. This was the first party of white men to view the Grand Canyon from the water.

Powell's account of his trip through the Grand Canyon is included in his book *Explorations of the Colorado River of the West and Its Tributaries*, first published in 1875. Powell, a geologist, also published the first classification of Native American languages.

Lieutenant John C. Frémont 1842–1844

The purpose of Frémont's first expedition was to explore the region between the Missouri River and the Rocky Mountains. Frémont and his party left Westport, Missouri, on June 1, 1842. The expedition followed the Kansas River for a distance, and then crossed over to the Platte River in Nebraska. The expedition party continued on into western Wyoming to explore the headwaters of the Green River. They then turned eastward to explore the Wind River Mountains and climbed Frémont Peak. Soon afterward, Frémont began the journey east by way of the Platte River to its outlet on the Missouri River. He reached Westport, Missouri, on October 1, 1842.

Frémont's second expedition party left Westport in May 1843, and traveled northwestward across Nebraska toward the Oregon Trail. Frémont wanted to find a route through the Rockies further south than South Pass in southern Wyoming. After having no luck in finding a new pass, Frémont turned around and continued on to the Great Salt Lake in Utah.

The expedition then went into the Wasatch mountains of Utah, then turned northward into what is now Idaho. They followed the Snake River north to the Columbia River in Washington. The men then traveled south along the Cascade Mountains of Washington and Oregon, and entered northwestern Nevada. They crossed the Sierra Nevada Mountains into northern California. After their explorations in California, Frémont and his men turned east back through the Sierra Nevada Mountains. From here, the men returned to Westport in July 1844.

Meriwether Lewis and William Clark 1804–1806

Captain Meriwether Lewis and Lieutenant William Clark led an expedition commissioned by President Thomas Jefferson. Their goal was to search for an easy water route across the continent. The expedition left St. Louis in May 1804, traveling upstream on the Missouri River to what is now North Dakota, and then taking canoes to the Rocky Mountains. When Lewis and Clark reached country where the rivers were no longer navigable, the men traded goods for much needed horses. The expedition continued on horseback over the Continental Divide and into northern Idaho. The party then traveled down the Clearwater River to the Columbia River in Washington. The river took them all the way to the Pacific Ocean in November 1805.

After reaching the west coast, Clark and most of his men turned around and headed east to explore the Yellowstone River in southern Montana. They returned to St. Louis in September 1806 on the Missouri River.

Source: *Who Was Who in World Exploration,* C. Waldman and Alex Wexler, 1992.

The Movement of the Steel Industry in the United States

Steel manufacturing required iron ore, limestone, coal, and the transportation to ship these materials, as well as the finished product.

1860–1910	Water transportation was still more common than rail during this period. Water provided a source of power and a means of transporting the steel. Pittsburgh's central location in a coal-rich area and the presence of three converging rivers made it a perfect location for the manufacturing of steel.
1910–1920s	It was discovered that large amounts of steel could be made more quickly by blowing a blast of air through molten iron. This process was called the Bessemer process. It required more coal than iron ore, so the steel industry was located close to the Appalachian Mountains coal supply. Coal was shipped to steel mills by trains. Iron ore deposits in the Mesabi Range in Minnesota were sent east by ships crossing the Great Lakes. As a result, the location of the steel industry shifted from Pittsburgh to the Great Lakes cities of Buffalo, New York; Erie, Pennsylvania; Cleveland, Ohio; and Toledo, Ohio.

U.S. Steel Production 1860–1980
in millions of tons

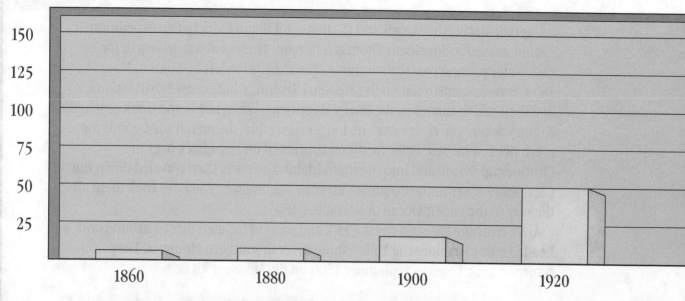

1920s–1940s	In 1923 John Tytus invented a process called continuous hot strip rolling that allowed massive sheets of metal to be produced. This technological breakthrough in the steel production process required more iron than coal. As a result the steel industry moved closer to the iron ore supply. Chicago, Illinois, and Gary, Indiana, had lower iron transport costs because they were closer to the Mesabi Range than Pittsburgh and Cleveland. Factory layouts were designed to be more efficient, and all components of steelmaking were centralized in a single facility.
1940s–1970s	As soon as an iron ore deposit was depleted, new locations were sought. As the iron ore deposits of the Mesabi Range became depleted, the steel industry shifted to the east and west coasts of the United States. Iron ore was then imported from Venezuela, Brazil, and Canada and transported by ship.
1970s–1990s	Due to mass destruction during World War II, Japan and Germany had to rebuild their steel mills. They did this using updated technology while the United States did not upgrade their mills. This, and high labor costs (average steel workers' wages are 2 to 10 times higher than in other countries) in the United States, enabled other countries to produce steel more cheaply than the United States steel industry. American iron and steel products are not competitive on the world market. As a result, the demand for American steelmaking has declined greatly. Many plants have closed down. What was once the great northeast manufacturing belt has become known as the Rust Belt.

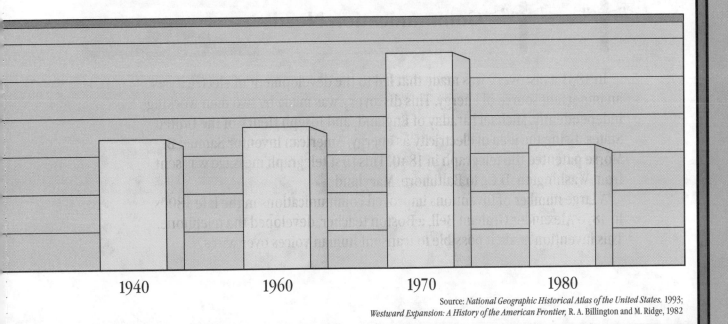

1940   1960   1970   1980

Source: *National Geographic Historical Atlas of the United States.* 1993;
Westward Expansion: A History of the American Frontier, R. A. Billington and M. Ridge, 1982

Advances in Technology Based on Different Needs

 ## Economic Needs

The Industrial Revolution (1700s–1800s) encouraged people to invent new machines to keep up with growing industries. One of the most important machines invented during this time was the steam engine. It was developed by James Watt in the 1760s. Watt's engine quickly became the chief source of power for transportation and industry.

While some people were involved in manufacturing, others were involved in agriculture, which had its own new list of inventions. An important invention was the reaper, a machine used to cut and collect grain. Cyrus McCormick, an American inventor, patented the reaper in 1834. Not only did grain have to be collected and cut, it also had to be separated from the stalk. A machine to do this, the threshing machine, was invented in 1834 by two American brothers, Hiram and John Pitts. Their machine became the model for modern threshing machines.

Another important agricultural invention was barbed wire, developed in DeKalb, Illinois, by Joseph E. Glidden of the United States in 1873. Barbed wire is a set of twisted wires with sharp hooks attached. It was often used as a fence to keep wild animals and unwanted people out of a farmer's pasture and away from his animals.

There were many more industrial and agricultural inventions that developed as economic needs changed. The inventions described above were major improvements and were widely used.

 ## Communication Needs

In 1831 a discovery was made that led to the development of electricity as an important source of energy. This discovery was made by two men working independently, Michael Faraday of England, and Joseph Henry of the United States. Using the idea of electricity as energy, American inventor Samuel B. Morse patented the telegraph in 1840. This first telegraph message was sent from Washington, D.C., to Baltimore, Maryland.

A large number of inventions improved communications in the late 1800s. In 1876 Alexander Graham Bell, a Boston teacher, developed the telephone. This invention made it possible to transmit human voices over wires.

Other inventions of the electronic age include the television and computers. A working model of the television was developed in 1926 by Scottish engineer John Logie Baird. In 1936 the British Broadcasting System transmitted the world's first open-circuit TV broadcasts.

More recently, developments in computer technology in the late 1980s and early 1990s have enabled people all over the world to communicate with one another, and to retrieve information from countless sources. The Internet, developed in Washington, D.C., is a system of computers connected by telephone lines and satellites. It allows people to send and receive messages and information from all over the world. The CD-ROM, developed in Redmond, Washington, is a compact disc that contains information that can be read by a computer. It stores large amounts of information on a small disc.

Improvements are always being made in televisions, computers, and other electronic goods. Inventors combine new ideas with existing ideas to create goods to keep up with our ever-changing communications needs.

 ## Transportation Needs

As industries and populations grew during the Industrial Revolution, the problem of transporting materials and people led to several new inventions. Robert Fulton invented the steamboat, and in 1807, the first successful steamboat service started in New York City.

Later, the gasoline powered engine invented by James Watt became important to further developments in transportation. In 1903 in Kitty Hawk, North Carolina, Orville and Wilbur Wright used a gasoline powered engine to power a small airplane they had built. The Wright brothers' plane became the first one to lift people into the air and fly successfully. Their invention served as a model for future airplanes and jets.

In 1957 the space age began when the former Soviet Union launched the first artificial satellite to orbit the Earth. John H. Glenn, a United States astronaut, orbited Earth in 1962. By the end of the 1960s, men were traveling to the moon. Humans first walked on the moon on July 20, 1969. United States astronauts Neil Armstrong and Edwin E. Aldrin traveled from the Apollo 11 spacecraft to the surface of the moon in a lunar module. In 1981 United States astronauts traveled through space in the first space shuttle. The space shuttle was developed at Cape Canaveral, Florida, and was the first manned spacecraft designed to be reusable.

Improvements in transportation came about as populations grew and more people needed to move from one place to another. Developments in space travel continue because we want to explore the vast, and largely unknown reaches of space.

GLOSSARY

artificial boundaries: boundaries formed when people choose a place to draw a line

astrolabe: an instrument used to measure the height above the sea of the celestial bodies such as planets and stars

boundary: an imaginary line between two states or countries

butte: an isolated, flat-topped mountain, or hill

cape: a point of land that extends into the water

caravel: small fast sailing ships developed in Portugal and Spain during the fifteenth century

cliff: a high, steep face of a rock

compass: shows which way north, south, east, and west are on a map

continents: one of the seven land areas of Earth, including Asia, North America, South America, Africa, Antarctica, Europe, Australia

cross-staff: an instrument used to find latitude by measuring the height of the sun and stars above the horizon

cyberspace: a computer network by which people can send messages to each other

degrees: a unit used to measure distance and location on Earth's surface

delta: an area of land formed by sediment at the mouth of a river

dune: a mound, hill, or ridge of sand that is formed by wind

elevation: altitude above sea level

equator: an imaginary line encircling Earth halfway between the North and South poles; degrees of latitude are measured from it

erosion: a process that strips rock and soil away from the surface of Earth and moves them to another place

fall line: a line on rivers where waterfalls and rapids begin

fault: a break in Earth's crust along which rock masses have been moved with respect to one another

glacier: a huge sheet of ice

globe: a model of Earth that shows the correct shapes of Earth's major land masses and bodies of water

historical map: a physical or political map that shows borders, physical features, or events of the past

immigrant: a person who settles in another country

irrigation: a way of supplying water through pipes, streams, and channels

labels: words/phrases that identify the symbols on a map

lagoon: a shallow body of water partly cut off from the sea by a strip of land

large-scale map: a map that shows much detail

latitude lines: horizontal lines north and south of the equator

legend (or key): a list of the map's symbols and what they represent

longitude lines (or meridians): vertical lines running from north to south on the globe

main image: a place/region shown on a map

map: a flat drawing of Earth

map scale: shows how many miles/ kilometers the inches/centimeters on a map represent

mental map: a map which represents the mental image a person has of an area

mesa: a large, flat-topped hill or mountain with steep sides surrounded by a plain

minutes: small units for measuring distance and location on Earth's surface; there are 60 minutes in one degree

mouth: the place where a river empties into a larger body of water

natural boundaries: boundaries formed by physical features such as rivers

physical map: a map that shows physical features such as mountains, rivers, oceans, and deserts

Piedmont: the foothills area along the southern Atlantic coast

plantation: a large farm that is devoted to a single crop and cultivated by workers who live on it

plate tectonics: Earth's outer crust is made up of huge slabs of rock called plates; these plates move about below Earth's surface in ways called faulting and folding; this process is called plate tectonics

plates: one of the huge sections that make up the land masses of Earth and the ocean floors; these allow for continents to move apart and come together in the plate tectonic theory

political map: a map that shows borders of states and countries and the location of capitals and other cities and towns

population map: a map that shows how many people live in a place

prime meridian: the starting point to measure both east and west around the globe; located in Greenwich, England

pull factor: something from a new place that attracts or pulls people away from their native country

push factor: something that encourages people to leave their native country

quadrant: an instrument used in navigation and astronomy to measure distance above the horizon

reef: a ridge of sand, rock, or coral that lies near the surface of a body of water

relief map: a map that shows high points and low points of a land area

reservoir: a natural or human-made place used for the storage of water

road map: a map that show cities, towns, places of interest, physical features, and the roads and highways that link them

small-scale map: a map that shows little detail

sound: a long, narrow body of water that runs parallel to the coast

sphere: an object shaped like a ball

strait: a narrow waterway or channel connecting two larger bodies of water

title: describes place/region shown on map

transcontinental railroad: a railroad that crosses a whole continent

tributary: a river or stream that flows into a larger river or stream

urbanization: the movement to, and clustering of, people in cities

waterfall: a flow of water from a high place over a ledge

weathering: the process of wearing down or change that occurs from being exposed to harsh climate

INDEX

A. Using the diagram, list the different forms of land and water in the appropriate columns below.

Landforms	Water Forms
butte	lagoon
cape	mouth
cliff	reservoir
delta	sound
dune	strait
mesa	tributary
reef	waterfall

B. Identify the landforms and water forms described below and complete the puzzle.

1. A stream that flows over the edge of a cliff. w a t e r f a l l
2. A narrow body of water that connects two larger bodies of water. s t r a i t
3. A river or stream that flows into a larger river or stream. t r i b u t a r y
4. The place where a river empties into a larger body of water. m o u t h
5. A point of land that extends into the water. c a p e

C. The following landforms and water forms are also on the map, but are not labeled. Label them. (Make sure students correctly identify the features listed below.)

1. mountain   6. volcano
2. river   7. glacier
3. ocean   8. bay
4. plateau   9. canal
5. hill   10. valley

D. Turn to the world map in the Almanac. How many other forms of land and water can you find? (Students might identify lake, gulf, island, peninsula, and so on.)

Lesson 1
ACTIVITY
Identify where you live on Earth and the forms of land and water in your community.

The Right Spot

The **BIG** Geographic Question

Where do you live and what landforms and water forms are found there?

In the article you were reminded of Earth's seven continents and four oceans. In the map skills lesson you learned to identify some of the landforms and water forms on Earth. Find out which of these landforms and water forms are in the area where you live.

A. Tell what you know about where you live.

1. In what continent and country do you live?

North America; United States

2. What is the name of your state?

(Make sure students identify the state in which they live.)

3. What is the name of your community?

(Make sure students identify the city, town, or neighborhood in which they live.)

B. Think about what your state looks like. (Students' answers should reflect characteristics of their state.)

1. Does your state have many lakes? _____

2. Is it near one of the four oceans? _____

3. Is your city nestled in a valley or located high in the mountains? _____

4. Write what you know about your state's landforms and water forms.

_____
_____
_____

C. Now think about your community and answer the following questions.

1. Is your community a city, a town, a village, or a suburb?

(Students' answers should reflect characteristics of their community.)

2. How many people do you think live in your community?

(Students' answers may be estimates of their community's population.)

3. Put a check beside each of the landforms and water forms listed below that are located in or near your community. Add to the list if necessary.

____ lake		____ waterfall
____ river		____ island
____ mountain		____ hill
____ plain		____ valley
____ plateau		____ cape
____ canal		____ bay
____ ocean		____ cliff
____ dam		____ coast
____ harbor		____ desert

4. Draw pictures or use construction paper or magazines to cut out pictures of landforms or water forms found in or near your community. (Students' pictures should match what they checked on the above list.)

D. Make a mobile to show some of the physical features that are found where you live on Earth. Attach the drawn or cut-out pictures that you have collected to pieces of string cut at different lengths. Tie the string to a hanger to make a mobile. Compare your mobile to your classmates' mobiles. (Check to make sure students' mobiles indicate characteristics of where they live in terms of physical features.)

A. Look at the maps and then complete the chart below by writing a brief description of the information you see on each map. (Descriptions for Maps 1 and 2 should appropriately match the maps.)

Map#	Description
1	
2	

B. Read the following definitions of some of the different kinds of maps. Then write the name of the correct type of map next to the map numbers below.

Types of Maps and Their Definitions

- Political map–a map that shows borders of states and countries and the location of capitals and other cities or towns
- Physical map–a map that shows physical features, such as mountains, rivers, oceans, and deserts
- Historical map–a physical or political map that shows borders, physical features, or events of the past
- Population map–a map that shows how many people live in a place
- Road map–a map that shows the streets, roads, or highways that can be used to get from one place to another

Map#	Type of Map
1	population; political
2	historical; political; population

C. Look at the two other types of maps in the Almanac on page 115. What types of maps do you think they are? Describe or define each one.

Map#	Type of Map	Definition
1	road map	a map that shows streets, roads, or highways
2	physical map	a map that shows physical features such as mountains and rivers

D. Look at the maps on pages 8 and 9.

1. What type of map do you think the Illinois Country map on page 8 is?

historical; physical

2. What type of map do you think the New York Citizens map on page 9 is?

historical; road

Lesson 2
ACTIVITY
Make a map to communicate important information.

Make Your Own Map

The **BIG** Geographic Question

What are the main features needed to create an effective map?

From the article you learned about the main features of a map. The map skills lesson helped you identify several different kinds of maps. Now make a map to communicate information to a friend or family member.

A. Under each of these headings, write titles for a map you might be interested in creating. Then put an X in the box next to the map you will create.

(Make sure students have indicated selections.)

☐ HISTORY (family, state, city)
Example: How My Family Came to the United States

☐ EVENTS (vacations, sports events, concerts)

☐ PLACES (school, home, neighborhood, city, country)

B. Answer the following questions about the map you are interested in creating.
(Students' answers should reflect an understanding of main map features and of political, physical, historical,

1. What type of map will best communicate information about your topic? _____ and road maps.)

2. What title best summarizes the map? _____

3. What main image will the map show? _____

4. What labels will be included on the map? _____

5. What information will need to be included in the map's legend? _____

6. Does your map need a scale? _____

C. Now that you have come up with a written list of things to include on your map, decide how it will look.

1. In the space below, sketch the symbols you will include in the legend on your map. (The legend should include all of the symbols that will appear on the map, along with a word or phrase telling what each symbol represents.)

2. Decide what the main image on your map will look like. Use the space below to work out a design for the map. Don't forget to include a compass if your map needs one. (Check students' sketches to make sure they are using map features effectively.)

D. Now that you have organized the parts of your map and made a sketch, create your final map on a separate piece of paper. When you have completed your map, show it to a family member, classmate, or friend. To see whether your map communicates what you wanted it to, have the person you share it with describe what they think the map is about.
(Maps should include a title, main image, labels, compass, legend, scale, and any additional graphics students sketched.)

A. Look at the map of North America and answer the following questions.

1. What are the three countries of North America?

Canada

United States

Mexico

2. What name is used to refer to the five lakes that form a boundary between

Canada and the United States? _Great Lakes_

Is it a natural or artificial boundary? _natural_

3. Name the river that forms a boundary between Mexico and the

United States. _Rio Grande_

Is it a natural or artificial boundary? _natural_

4. Name the area of land that is on the northwestern boundary of Canada and the

country that owns that land. _Alaska; the United States_

Is this a natural or artificial boundary? _artificial_

5. What is the number of the parallel line that forms a long section of the

boundary between Canada and the United States? _the 48th parallel_

Is this a natural or artificial boundary? _artificial_

**B. Look at the coastline boundaries of North America and complete the
following questions.**

1. What ocean borders the east side of the United States and Canada? _Atlantic Ocean_

2. What ocean borders the west side of North America? _Pacific Ocean_

3. What ocean is north of Canada? _Arctic Ocean_

4. What large body of water lies between Mexico and the Atlantic Ocean? _Gulf of Mexico_

C. Draw some conclusions about the above information.

1. What does the map of North America tell you about boundaries? _(Possible answer: Most of them_
seem to be natural boundaries like rivers and lakes; however, there are some artificial boundaries, too.)

2. Why do you think waterways are common boundaries in North America? _People have historically_
settled near rivers, lakes, and streams that could be used for drinking water or transportation.

Lesson 3

Find out what North America looked like
before it was first settled by people.

North America Long Ago

The **BIG**
Geographic Question

How has the landscape of North
America changed as a result of
the last Ice Age?

From the article you learned how forces of nature shaped Earth. The
map skills lesson showed you some of the natural boundaries of North
America that these forces formed. Now find out how some of the physical
characteristics that resulted from the last Ice Age have created big changes
in North America's vegetation.

**A. Take notes about what North America was like after the last Ice Age.
Use the physical map of North America in the Almanac to find this
information.**

1. **Land Notes**—Many of the landforms we see today were created as ice
moved across North America. Glaciers of long ago pushed and
pulled at the land to shape the mountains and plains of today.

a. Name the major mountain ranges of North America. _Rockies, Sierra Nevada, Appalachians_

b. Name the plains that are located along North America's eastern and

southern coastlines. _Atlantic Coastal Plain, Gulf Coastal Plain_

c. Name a major plateau in North America. _Colorado Plateau_

d. What is the name of the large stretch of plain in central North America? _Great Plains_

2. **Water Notes**—Several North American lakes and rivers were created as
glaciers melted and retreated north as the Ice Age was coming to an end.
Land and valleys carved by the moving ice filled with water as the ice
melted.

a. List the major lakes of North America. _Great Lakes_

b. What is the longest river in North America? _Mississippi_

---

c. What river runs along the boundary between the present-day countries of

Mexico and the United States? _Rio Grande_

**B. Now that you have learned how the Ice Age changed the physical landscape
of North America, use the vegetation maps in the Almanac to answer
questions about vegetation patterns before and after the Ice Age.**

1. **Vegetation Notes Before**—Vegetation patterns were much different before
the moving glaciers and melting ice changed the physical features of
North America. (Students' answers should reflect the patterns shown
on the map on page 114 of the Almanac.)
Where were the following vegetation regions located at the end of the last Ice Age?

a. deciduous forests _____

b. coniferous forests _____

c. grasslands _____

d. mixed forests _____

2. **Vegetation Notes Today**—Patterns in the growth of vegetation in North
America have changed as a result of the Ice Age. The presence of rivers
has provided enough moisture for new types of vegetation to grow. Till,
the fertile soil left behind as mountains and plains were formed, also
allowed new types of vegetation to grow. (Students' answers should reflect the patterns shown
on the map on page 114 of the Almanac.)
Where are the following vegetation regions located today?

a. deciduous forests _____

b. coniferous forests _____

c. grasslands _____

d. mixed forests _____

**C. Based on all of the above information, explain the growth of present-day
vegetation in North America. Think about how plants need fertile soil and
moisture to grow. Include information on how the Colorado and Mississippi
rivers have contributed to changes in vegetation.**

**A. Study the map of North America. Circle and label the following regions
of the United States.** (Make sure students have correctly identified the following regions on the map.)

1. Northwest
2. Southwest
3. Eastern Woodlands
4. Great Plains

**B. Compare the color patterns for high and low population densities on
the map.**

1. Which color represents a low population density? _light yellow and gold_

2. Which color represents a high population density? _brown_

3. Which areas of the United States had a low population density? _Great Plains, Midwest, and_
parts of the Southwest

4. Which areas of the United States had a high population density? _California and parts of the_
Northwest

**C. Use the map, the information on pages 20–21, and the Almanac to
describe the physical features of the following areas.**

1. Northwest _forests, ocean, mountains, trees_

2. Southwest _deserts, mountains, canyons, desert clay_

3. Eastern Woodlands _trees, rivers, lakes, swamps_

4. Great Plains _grasslands, rich river valleys_

**D. Explain why some areas in the United States might have had an especially
high population density. Base your answer on what you know about the
geography and natural resources of the four regions prior to the time
of Columbus.**

(Students' answers should express the following idea: Areas with a lot of different
food sources, such as fish and game, and plenty of materials for building might explain
large communities being located there; whereas areas with only one or two food sources,
such as the bison or a few plants and animals, and inadequate materials for building
wouldn't support many communities. Also areas located near large bodies of water might
have higher populations because the water could be used for drinking and transportation.)

Lesson 4

ACTIVITY Create alternate dwellings for Native American tribes.

Native American Dwellings

The **BIG**
Geographic Question

What alternate dwellings could Native
American tribes of long ago have built
based on the geography and resources
of the regions in which they lived?

From the article you learned how Native Americans of long ago adapted
their ways of living to the physical geography and natural resources around
them. The map skills lesson showed you where many or few Native Americans
of long ago lived. Now describe and draw or create a model of an alternate
dwelling for each of the following Native American tribes: the Makah, the
Navajo, the Cheyenne, and the Seneca.

**A. Use the Almanac to find the region in which each Native American
tribe below lived.**

1. Makah _Northwest_   3. Cheyenne _Great Plains_

2. Navajo _Southwest_   4. Seneca _Eastern Woodlands_

B. Use the article and Almanac to answer the following questions.

1. What were the physical features of the region in which each group lived?

Northwest: mountains, forest, coastal area; Southwest: deserts, mountains, canyons

Great Plains: grasslands, river valleys; Eastern Woodlands: forests, mountains, swamps, coastal area

2. What natural resources were available to each group?

Northwest: trees, whales, fish, game; Southwest: desert clay, soil for growing corn and cotton

Great Plains: rich soil, buffalo; Eastern Woodlands: trees, game, fish

3. What was the climate like in each region?

Northwest: mild climate, heavy rainfall; Southwest: hot and dry, little rain

Great Plains: warm summers, cold winters; Eastern Woodlands: warm summers, cold snowy winters

---

**C. Organize the information you have gathered onto the chart below.
Use the article and Almanac to find any missing information.**

Tribe	Region	Climate	Use of Resources	Types of Past Shelter	Drawing of Shelter
Makah	Northwest	heavy rainfall and mild temperatures due to Pacific Ocean air currents	trees for housing and canoes; clothing woven from cedar bark	plank houses	
Navajo	Southwest	hot and dry with little rainfall	soil for growing corn, cotton, and beans along with providing raw materials for housing	hogans	
Cheyenne	Great Plains	warm, dry summers; cold winters with blowing snow	buffalo skins for shelters, shoes, clothes	tepees	
Seneca	Eastern Woodland	warm summers; cold snowy winters	trees for housing and clothes	longhouses	

**D. Review the information on the above chart. Select one of the tribes and
design an alternate dwelling the people of the tribe could have built using
the resources they had. Explain how this structure would have been useful
in the environment (climate and geography) of the region. Sketch your
alternate dwelling on the chart. Make a model of the alternate dwelling.**
(Alternate dwellings may include the following: Makah: tents made of animal skins; Navajo:
caves in the mountains; Cheyenne: earth or sod houses; Seneca: tents made of animal skins)

A. Use the map scale to answer the following questions.

1. How many miles are represented on the scale?

200

2. Using a ruler and the map scale to determine the distance, approximately how many miles by air is Boston from New York City?

approximately 200 miles

B. Use a piece of string and a ruler to measure the distance by water between Boston and Cape Hatteras on the map and answer the following questions.

1. How many inches is the length of string that represents the distance from Boston to Cape Hatteras? _____3_____ inch(es)

2. Measure the miles line on the map scale. Complete the following.

_____1_____ inch(es) is equal to _____200_____ miles.

3. How many miles by water is Boston from Cape Hatteras? approximately 600 miles

C. Use a string and the map scale to measure and determine the distance of the land and water routes between Boston and Houston.

1. How long is the land route from Boston to Houston? approximately 1,875 miles

2. How long is the water route from Boston to Houston? approximately 2,500 miles

3. If you had been traveling in the 1400s from what is now Boston to what is now Houston, which route would you have taken? Think about the ways people traveled during that time. Explain your answer.

(Students' answers should reflect their understanding that the water route would have been faster

because efficient ships were available. There were no roads.)

4. If you were traveling from Boston to Houston in the 1990s, would you take the water or land route? Explain why.

(Students' answers might be that they would take the land route because there are good roads and

good automobiles. The water route would be possible, but there is little passenger travel along this

route. It would be expensive and time consuming.)

Lesson 5

ACTIVITY   Find out how travel time, distance, and technology are related.

How Far? How Long?

The **BIG** Geographic Question

How is the time it takes to cover a particular distance related to the transportation technology available?

In the article you learned what a difference technology made to explorers in the fifteenth century. The map skills lesson showed you how to use a map scale to locate places around the world using lines of latitude and longitude. Now find out how the means of transportation have continued to advance. How has technology helped people go farther in less time!

A. Gather information from the Almanac to complete the chart below.

Traveler and Date of Journey	From/To	Method of Travel	Distance	Time
Marco Polo (1271–1274)	Venice– Shangdu, China	boat and camel	7,500 miles	3 years
Christopher Columbus (1492)	Spain– San Salvador	caravel	4,500 miles	2 months, and 9 days
Ferdinand Magellan (1519–1521)	Spain, around South America, back to Spain	caravel	50,000 miles	3 years
Meriwether Lewis and William Clark (1804–1806)	St. Louis– Pacific Ocean	foot, horse, and boat	1,600 miles	18 months
Charles Lindbergh (1927)	New York– Paris	airplane	3,610 miles	33 1/2 hours
Modern air travelers (1996)	New York– London	supersonic jet	3,469 miles	3 hours

---

B. Using the information from the chart, answer the following questions.

1. What was the fastest form of long-distance transportation in the fifteenth and sixteenth centuries? sea travel or caravel (boat)

2. What is the fastest form of long-distance transportation today? air travel or plane

3. In what century did the greatest advances occur in traveling long distances in a short time? What method of travel made these advances possible? 1927 (twentieth century); air travel

4. Which of the forms of transportation listed on the chart will probably continue to get faster as technology advances? air travel

5. Use a calculator to figure out how many miles per hour Charles Lindbergh covered on his trip from New York to Paris. Compare that figure with a flight in 1996 from New York to London by supersonic jet. List the figures below.

Charles Lindbergh's miles per hour 108

Modern air travelers' miles per hour 1,156

C. Complete the following regarding future travel and tools.

1. The people you studied traveled over land, across water, and through the air. Today people are exploring other parts of our world in various ways. List some places that are still being explored.

(Students may list mountains, oceans, the Arctic, the Antarctic, and space.)

2. How do you think people will travel to these places that are still being explored?

(Students' answers might include the use of four-wheel-drive vehicles, helicopters,

submarines, dog sleds, ice-breaking ships, undersea robots, and space shuttles.)

3. What do you think people will find in these new places? For example, what do you think explorers would find in space that would improve life on Earth?

(Possible answers include: Explorers in space may find water sources on other planets

and information about the atmosphere and the ozone layer. This information might help

people on Earth understand how long humans can live in space and help find new resources.)

A. Look at the map of Columbus' travels in 1492 and answer the following questions.

1. Columbus left Palos, Spain, on August 3, 1492. What is the approximate latitude of Palos, Spain?

39°N

2. What is the approximate longitude of Palos, Spain? 7°W

3. What are the approximate latitude and longitude of the Canary Islands, where Columbus stopped in August 1492?

29°N 16°W

4. If Columbus was at 30°N latitude and 60°W longitude, where would he be?

in the Atlantic Ocean

5. At about what latitude was Columbus when he landed on San Salvador?

24°N

6. Just south of San Salvador is a line of latitude with a special name. It is the northern boundary of the tropical region. What is this line of latitude called?

Tropic of Cancer

7. If Columbus had landed at 30°N latitude on the North American continent, where would he have landed?

present-day Florida

B. Looking at the map, number the following events of Columbus' journey in the correct order. Beside each different place where Columbus stopped, write its latitude and longitude.

Sequence	Events	Latitude and Longitude
3	Columbus lands at San Salvador Island.	24°N, 74°W
4	The Santa Maria goes aground on a reef.	
6	Two ships return to Palos.	37°N, 7°W
5	The Niña and Pinta begin their return voyage.	
2	Columbus reaches the Canary Islands.	29°N, 16°W
1	Columbus sails three ships out of port Palos, Spain.	37°N, 7°W

Lesson 6

ACTIVITY   Find out why people explore new places.

Exploring New Places

The **BIG** Geographic Question

What motivates people to explore and to move to new places?

From the article you learned some reasons the sixteenth century became the great age of European exploration. The map skills lesson showed how to locate places around the world using lines of latitude and longitude. Now think about what would have made you want to explore if you had lived in the 1500s. What would make you want to explore now?

A. Push/pull factors are reasons that make one decide to travel to a new place, or stay at home. **Push factors** are influences at home that encourage people to move away. **Pull factors** are influences in a new place that encourage people to move there.

1. Think about what you read in the article. Make a list of sixteenth century push/pull factors.

Push Factors	Pull Factors
lack of farmland	adventure
escape religious persecution	religious freedom
poverty	chance to become wealthy
hunger	abundant land
	jobs

2. Select one character to represent from the following list of people who were interested in exploration in the sixteenth century. Circle your choice. (Make sure students circle one character.)

king/queen	merchant	mapmaker
European citizen	explorer	sailor

B. Using the idea of push/pull factors, think about why you, a sixteenth-century individual, want to leave your home and explore a faraway, unknown place. Or consider why you want to stay home and support a trip made by an explorer. Think about why you want to explore or why you want to stay at home. Do you have a marketable talent or skill? Do you have resources for a voyage of exploration? Write a couple of sentences that describe your motivation and your talent.

(Students' sentences should include relevant details for their characters. For example, "I, Queen

Isabella of Spain, would love to explore new places, but since I must stay home to rule my country,

I will give my money and encouragement to brave sailors.")

C. The great age of European discovery took place almost 500 years ago. Are there still places to explore? If so, where?

(Students may mention outer space, mountains, oceans, and the Arctic and Antarctic areas.)

D. Think about these places to explore today.

1. Make a list of the push/pull factors that might be related to this present-day exploration.

Push Factors	Pull Factors

(Make sure students' answers reflect an awareness of today's exploration and advances in technology and that the push/pull factors they listed make sense in the context of the exploration they indicated.)

2. Select one of the following characters who might be interested in exploration today. Circle your choice. (Make sure students circle one character.)

The President	merchant	scientist
American citizen	explorer	geographer

3. Think about how you will be able to participate in or support the exploration. Do you have a marketable talent or skill? Do you have resources that would be helpful to exploration? Write a couple of sentences that describe your motivation and talent or resources.

(Students' sentences should make sense in the context of the exploration they indicated. They should be specific and creative. For example, "I am an ocean explorer who wants to find a way for people to live far beneath the water. I will participate in the exploration and share knowledge of the underwater world.")

E. Discuss your sentences with a classmate or friend and get feedback from them. Write on a separate sheet of paper a rap song, skit, poem, or story about your exploration. Present your work to the class.

(Students' presentations should be creative and clearly communicate information about the explorer and exploration they selected.)

Lesson 7
MAP SKILLS
Using a Map to Look at Resources of Early Settlements

The Spanish, English, and French settled areas of the present-day United States. They found and used resources that would allow them to make a living trading with each other and with those who stayed in Europe. Look at the resources that made trade possible among these groups.

Use the map to complete the following.

1. Shade the regions on the map where people from the following countries settled. (Students should shade the regions as shown on the map on page 39.)

   a. England   b. France   c. Spain

Map Key

Cattle	Wheat
Swine	Corn
Sheep	Indigo
Wood products	Iron Forging
Tobacco	Gold / silver mining
Hemp	Furs / animal skins
Rice	

Resources of the 1700s

---

2. Name two farm products raised in the 1700s by settlers from each of the following countries. (Students' answers might include the following.)

   a. England   tobacco, cattle, corn, wheat, swine, sheep, rice

   b. France   cattle, corn, hemp, rice, swine

   c. Spain   cattle, sheep, corn

3. Some of the resources the settlers found did not have to be grown, but still came from the land. These types of resources are raw materials. Name two kinds of raw materials found in the 1700s by settlers from each of the following countries. (Students' answers might include the following.)

   a. England   wood products, iron

   b. France   iron, furs, indigo

   c. Spain   wood products, indigo, gold and silver

4. List below each of the products you named above and how Europeans and Native Americans might have used these farm products. (Make sure students' descriptions of each product's use correspond with the product.)

Farm Products	European Uses	Native American Uses

5. Which group of early colonists relied most heavily on farming? _the English_

6. Which group of early settlers relied most heavily on the land's raw materials? _the French_

7. Do you think the Spanish, French, and English settlers used more of the products they raised in North America or traded more with their native countries? Explain your answer.

   (Students' answers should indicate that the Europeans traded more of their products with their native countries for profit than they did with each other in North America.)

Lesson 7
ACTIVITY
Find out about some early North American settlement types and locations.

Building a Settlement

The BIG Geographic Question

What types of settlements did groups from European nations create in North America and where were they?

From the article you learned that the English, the French, and the Spanish came to North America for different reasons and settled in different regions. In the map skills lesson you looked at some of the products the Europeans raised and how these products were used. Decide where you would have settled in North America in the 1700s.

A. Use the Almanac to identify each settlement's purpose and features of the land areas where they originated and were settled. Complete the chart below.

European Group	Type of Settlement	Purpose	Land and Climate Features in Europe	Land and Climate Features in North America
English and French	fort	built to protect against Native American attacks	mountains and plateaus in the north; gently rolling plains and hills in the central and southern regions; mild climate with ocean winds and plentiful rain	mountains and fertile pastures; bays, inlets, and islands along coastline; cold winters and warm summers
French	trading post	built as a base for trading furs with Native Americans	flat or rolling plains; hills and mountains in the east and south; warm summers and cool winters; warm throughout the year on southern coast	rough, rocky plateau in the north; plains in the west and south; mountains in the southeast; many rivers; long, cold winters and short, warm summers
Spanish	mission	built to spread Catholic religion to Native Americans	plains broken by hills and low mountains; poor soil for growing crops; dry, sunny weather most of the year	dry deserts dotted with rock formations; deep canyons, flat plains, and grasslands; hot summers and warm, pleasant winters

---

B. Select one of the following to design: a Spanish mission in San Antonio, Texas; a French trading post in New Orleans, Louisiana; or an English fort in Philadelphia, Pennsylvania. Write your choice on the line below.
(Make sure students indicate their choice of the type of settlement they would like to build.)

C. Look at the map on page 40. Mark the location of your settlement choice. (Make sure students mark the correct location of San Antonio, New Orleans, or Philadelphia on the map on page 40.)

D. To help plan your settlement, find out about the following physical features of the area where it is located. Look in the Almanac for help with this information. (Students should list the appropriate physical features of the location of their settlement choice.)

   1. climate _____

   2. vegetation _____

   3. waterways _____

E. Draw the design of your settlement choice in the space below. List off to the side of your drawing the area's natural resources that you used in planning your settlement's construction. On the lines below, explain how the North American terrain and the settlement's future dealings with the mother country influenced its location and affected your design.

(Students should use their imaginations and what they know about settlements to make their designs. The list of natural resources should include only those that are indigenous to the region.)

Explanation of Settlement: (Explanations should reveal an understanding of how location, and its similarity to the mother country, and the settlers' purpose for coming to North America influenced the type of settlement.)

A. Look at the map and its key. Find elevations in the following regions of the United States.

1. Rocky Mountains 5,000 feet and above   3. Coastal Plain 0 to 1,000 feet
2. Great Plains 2,000 to 5,000 feet   4. Central Lowlands 1,000 to 2,000 feet

B. Use the map to help you answer the following questions.

1. French explorers were mainly interested in fur pelts rather than settling land.

   a. Through what areas did the French trade routes extend?
   the northeast, Great Lakes region, and the Appalachian Mountains

   b. What geographical features might have presented difficulties for the French traders? Why? They may have had difficulty crossing the Appalachian Mountains because of their
   elevation and the Great Lakes because they are large bodies of water.

   c. Which Native American tribes might have interacted with the fur traders?
   Miami, Huron, and Iroquois

2. The English settlers were interested mainly in raw materials and food to send back to England.

   a. Which area included most of the English territory in the United States?
   the East Coast

   b. Which Native American tribes might have taught English settlers how to plant and grow crops? Powhatan and Iroquois

3. In what region did the Europeans introduce horses to Native Americans and why do you think horses became so valued in this region of the country?
   Great Plains; the land was flat, and horses could travel great distances in less
   time than it took Native Americans on foot

4. Which Native American tribes might have interacted with the Spanish missionaries?
   Pueblo, Hopi, and Navajo

Lesson 8

ACTIVITY Learn about the different views on how land should be used.

Debating for Land

The **BIG** Geographic Question   How should land be used, and who should own it?

From the article you learned about conflicting Native American and European ideas about using land. The map skills lesson showed you various Native American tribes and physical features of the regions in which they lived when contact was first made with Europeans. Now conduct a debate as you negotiate, or bargain, for land.

A. Using information in the article, name four key geographic regions in North America at the time of early European settlements. (Answers should include regions from the article or the map.)

1. Appalachian and Great Lakes   3. Great Plains
2. Eastern Woodlands   4. Southwest

B. Choose one of the regions that you listed and answer the following questions about that region. (Students' answers should be appropriate to the chosen region.)

1. Which Native American tribes lived in this region?

2. What were the major resources in the region?

3. How did the Native Americans who lived in that region use the land?

C. The Native Americans, English farmers, Spanish missionaries, French traders, and representatives of European governments had different ideas about how to use North America's land and resources. On the chart below, write each group's ideas about how to use the land.

Group	How They Wanted to Use the Land
(Students should list historical groups appropriate to their chosen region. Descriptions of the groups' ideas should reflect an understanding of the various viewpoints on land use.)	

D. Evaluate each group's ideas about how to use the land and its resources and whether land should or should not be owned.

1. Choose a point of view to represent.

2. Write notes below to support your point of view and to prepare for a debate.
   (Students should give reasons that support their opinions about whether land should or should
   not be owned.)

A. Use the New York state map to answer the following questions.

1. Where are the areas of greatest population density in the state?
   The area around New York City and its suburbs, Albany, Syracuse, Rochester,
   and Buffalo

2. How would you account for the low population density in the light yellow?
   They are in the Appalachian Mountains.

3. What geographic feature might account for the population density represented by a strip of orange through the middle of the state?
   Population is usually higher along waterways such as rivers and lakes.

B. Use the map of New York City to answer the following questions.

1. What are the five boroughs of New York City?
   Staten Island, Manhattan, the Bronx, Queens, and Brooklyn

2. Which borough is the southernmost borough?
   Staten Island

3. Which boroughs are located on Long Island?
   Brooklyn and Queens

C. Compare the two maps and answer the following questions.

1. From which map would you get more detailed information, the city map or the state map?
   the city map

2. Which is the large-scale map?
   the city map

3. Which is the small-scale map?
   the state map

4. Which map would you use to compare the populations of several cities?
   the state map

5. On which map could you find more information about a specific area?
   the city map

Lesson 9

ACTIVITY Create a visual display representing New York City's population change over time.

New York City—Then and Now

The **BIG** Geographic Question   How can we show changes in population over time?

From the article you learned about the role of New York City as the first capital of our nation, a port of immigration and trade, and a center of business and culture. From the map skills lesson you learned how the city's population is distributed. Now discover how New York City's population has changed over time.

A. Look at the following table of New York City's population from 1750 through 1990.

1. Using the table, draw a bar graph showing the population for each year shown. The year 1750 has already been done for you.

2. Give your graph a title.

Title _____ (New York City Population, 1750–1990)

Year	Population
1750**	25,000
1800	60,000
1850	696,000
1900	3,400,000
1950	7,900,000
1990*	7,300,000

Figures are rounded.
*As of last U.S. Census
**Estimated

B. Use your graph and information from the article to answer the following questions.

1. What were some of the factors that probably led to the increase in New York City's population:

   a. from 1750 to 1800? Congress made New York the nation's capital, the Constitution was passed,
      and George Washington was inaugurated as the first president.

   b. from 1800 to 1850? New York became a center for international trade. Businesses were started.

   c. from 1850 to 1900? Many immigrants from Europe passed through the port of New York and settled
      in the city. Business, financial, and cultural opportunities continued to grow.

   d. from 1900 to 1950? Growth continued as people came from other parts of the world, and New
      York became a center of world trade.

2. Between 1950 and 1990, the population changed as follows:

1950	7,900,000
1960	7,800,000
1970	7,900,000
1980	7,100,000
1990	7,300,000

   a. Why do you think the population dropped between 1970 and 1980?
      (Students may speculate that population dropped between 1970 and 1980 because fewer
      immigrants came to the city due to overcrowding.)

   b. What might you expect the population to be at the next census in the year 2000? Explain your answer.
      (Students may suggest that the population will remain in the 7 millions because the city probably
      has about as many people as it can hold in the area. Also, the population has remained fairly
      steady for the last 50 years.)

C. Create a visual aid, such as a bar graph, pie chart, or table to show New York City's ethnic population mix in 1990. (The graphic should be divided as follows: a little over 50% white, 28.8% black, 23.7% Hispanic, 7% Asian,

1. Look in the Almanac to find the information you need. and .3% Native American, Inuit, Aleut.)

2. Organize your information by using the percentages to divide the 1990 bar of your graph on page 54 into sections.

3. Create your visual aid and label each section with the ethnic group it represents.

A. Use the map to answer the following questions about specific location.

1. What city lies at a latitude of 41°N and a longitude of 74°W?

   New York City

2. About how many degrees of latitude are there between Philadelphia and New York City? one

3. About how many degrees of longitude are there between White Plains, New York and York, Pennsylvania? about 4 degrees

4. The Schuylkill River lies between what lines of latitude? 40N–41N

B. Look at the entire area that the map shows and answer the following questions.

1. About how many degrees of longitude does the map cover? 13

2. About how many degrees of latitude does the map cover? 10

C. Using the map, identify the places described below and answer the following questions.

1. If point A is located at 42°N and 75°W and point B is located at 40°N and 74°W, in what direction would you travel to get from point A to point B?

   southeast

2. Is there a city at 39°N and 74°W? If so, what city is it?

   No, that location is in the Atlantic Ocean.

3. About how many degrees of latitude are there between Yorktown, Virginia, and Philadelphia, Pennsylvania?

   about 3 degrees

4. What are the latitude and longitude coordinates for Lexington, Massachusetts?

   42°N and 71°W

5. What are the latitude and longitude coordinates for York, Pennsylvania?

   40°N and 78°W

Lesson 10

ACTIVITY
Explore the effects of physical geography on various Revolutionary War battles.

Geography in Battle

The BIG Geographic Question
How did physical features affect the outcome of Revolutionary War battles?

From the article you learned about conditions at Valley Forge and why Washington chose that location for his army's winter camp. From the map skills lesson you learned about the location of some Revolutionary War battles. Now find out about three other Revolutionary War battles.

A. Use the Almanac to complete the following.

1. Identify and list the physical features surrounding the following battle sites.

   Yorktown: peninsula, James River, Chesapeake Bay, York River

   Lexington/Concord: hills, lakes, shallow rivers

   Trenton: Delaware River, Atlantic Coastal Plain, forests

2. In which battles was weather an influencing factor? Explain why. Yorktown: storms and heavy rains flooded the boats used in Cornwallis' attempted escape; Trenton: cold weather made the Delaware River too cold to cross, cold temperatures and sleet made it difficult for the troops to march forward

B. Use the Almanac to read more about the battle sites listed above. Make notes in the chart below. (Students' notes should include location and outcome of the battles, as well as the details surrounding each battle.)

Yorktown	Lexington/Concord	Trenton

60

C. Look back at the physical features you listed for each battle site.

1. Suppose you were the defending army at each battle location.

2. Put a + above the features that you think were positive location factors.

3. Put a – above the features that you think were negative location factors.

D. Compare your list of positive and negative factors to your notes from the Almanac. (Students' answers should show an understanding of how physical features can be positive or negative location factors.)

1. Did you come up with the same physical features and locational factors?

2. Were you accurate in identifying which physical features would be helpful to you as a defending army? Explain your answer.

E. Describe how the geographic setting of each battle site affected the outcome of the battle.

   Yorktown's location on a peninsula was the downfall of Cornwallis and his troops.

   American and French troops surrounded the British, leaving no way to escape.

   The geographic setting of Lexington/Concord itself had no direct effect on the outcome of the battle.

   Trenton's location on the Delaware River enabled Washington and his men to cross the bridge and then block it, preventing a British escape.

61

B. Use the map and the Almanac to help you answer the following questions.

1. Through what present-day states did the Oregon Trail pass?

   Missouri, Kansas, Nebraska, Wyoming, Idaho, Oregon

2. In what mountains is South Pass located? The Rocky Mountains

   How high is South Pass? 7,550 feet

3. What major rivers did the pioneers follow during their journey?

   North Platte River, South Platte River, Snake River, Columbia River

C. Use a piece of string, a ruler, the map, and the map scale to figure out the following times and distances.

1. What is the distance between Independence and Fort Laramie?

   about 650 miles

2. How long is the Oregon Trail from Independence to Oregon City?

   a little over 2,000 miles

3. If the wagon train averaged 15 miles per day, how long would it take them to make the journey from Independence to Oregon City?

   133 days (about 4½ months)

D. Look at the map and describe what a journey along the Oregon Trail would have been like.

1. Why do you think the trail mostly followed rivers?

   The rivers were a constant source of drinking water; and sometime a travel source when overland travel was difficult.

2. What geographic features would have slowed the settlers down? What features would have given them the greatest challenge? Explain your answers.

   Crossing the rivers and South Pass would have caused the greatest difficulty because of the problems of getting wagons and cattle through these areas.

65

Lesson 11

ACTIVITY
Write a diary entry describing how it might have felt to travel across the country in a covered wagon.

Dear Diary

The BIG Geographic Question
What physical challenges did the pioneers meet as they moved westward and how did they handle them?

From the article you learned what life was like for the pioneers who settled the West. In the map skills lesson you identified the physical features that the pioneers faced on the trail west. Now explore and describe how people felt and what they thought as they made their difficult journeys westward.

A. Select one of the trails shown on the map of westward expansion on page 62: the National Road, the Oregon Trail, the Mormon Trail, the California Trail, or the Santa Fe Trail. Answer the following questions. Use the Almanac to find out more about the trail you chose. (Students' answers to the following questions should include information about the geography of the trail chosen; Native American tribes encountered; weather, food, water, and fuel along the way; and difficulties encountered, such as illness or homesickness.)

1. How many miles long was the trail?

2. What mountain ranges, rivers, deserts, or other physical challenges did the people traveling it encounter?

3. What sources of food and water might the pioneers have had along the way?

4. Did the pioneers cross any land that was occupied by Native Americans? Explain your answer.

B. Imagine that you are a member of a pioneer family traveling on the trail you selected and keeping a diary. Complete the following to help organize the information that will go in your diary entries. On a separate piece of paper, write two or three entries describing some part of your trip. Date the entries in your diary with appropriate dates. (Students' answers should include the writer's feelings about and reactions to what he or she encountered on the trail chosen.)

1. Describe how you felt about leaving home.

2. Describe various physical features that you encountered during your journey westward.

3. Describe any hardships or challenges that your family faced and tell how you overcame them.

4. Describe what a typical day was like. (What did you eat? How did you cook? How many miles did you cover in a day? What did you use for fuel? What were your duties? Did you have time for play or other activities?)

5. Describe how you felt about the trip west by considering the following questions. Were you afraid or excited? Were you worried about making the journey? Did you find the conditions of the trip depressing, or were you optimistic and amazed by the things you saw?

6. Include a final entry about what it was like to reach your destination.

B. The map shows the routes of four famous expeditioners who went West. Using the map, answer the following questions.

1. Which explorer circled the land around the Great Basin?

Frémont

2. Which explorer traveled up the Arkansas River into the mountains, and then crossed the Rio Grande twice?

Pike

3. Choose one of the explorers listed on the map key. Write a description of the route he followed. What kinds of landforms did the explorer encounter?

(Students' descriptions should include information about physical features true to the

chosen explorer's route.)

C. Physical features can be used as boundaries when dividing land. Suppose you were a settler traveling westward in the mid-1800s.

1. Choose a destination for your trip.

(Make sure students choose a territory west of the Mississippi River.)

2. What route would you follow to arrive at your destination?

(Make sure students' answers include a description of a route to get to their destination.)

3. What kind of land features might you find along the way?

(Such features could include the rivers; mountains; desert valleys; and dry,

rugged plains that are indicated on the map.)

4. How might you use the land where you settle?

(Possible answers include: for farming, mining, trapping, raising cattle)

5. Your wagon train has a total of 30 wagons, each containing one family. Devise a method for dividing the land where you will settle. What boundaries will surround each family's plot of land? How will you decide what boundaries to use?

(Possible answers include: Boundaries may be determined by natural land features such as

rivers, mountains or lakes; land may be divided along natural boundaries or along lines determined

by law or agreement; land may be parceled out in square or rectangular lots.)

Lesson 12

ACTIVITY Create a work of art to represent a historical expedition.

Recreating History in Art

The BIG Geographic Question What was the scenery and daily life of the people in the American West like?

From the article you learned about several explorers of the West. The map skills lesson showed you the routes of their expeditions. Now imagine you are an artist who has been hired to go along on one of these trips and illustrate what the explorers saw on their journeys.

A. Choose one of the expeditions described in the article and charted in the map skills lesson and write the explorer's name below.

(Make sure students select one of the expeditions.)

B. Answer the following questions about the land that your expedition party will explore. Look at pages 70–71 and in the Almanac for help with this information.

1. Describe the route you will travel. (Students' answers will vary depending on the expedition they select, however, they should reflect information appropriate to the selection.)

2. Record information about the following: (Students' answers will vary depending on the expedition they select. However, they should reflect information appropriate to the selection.)

a. waterways

b. mountains

c. other landforms

3. Describe what you think the climate would be like in the areas as you traveled through them. Use what you know about the landforms and the time of year in which you are traveling.

(Students should describe climates appropriate to the time of year their expeditions took place and

physical features encountered along the way.)

C. Sketch a map below of the route you traveled. Be sure to indicate the landforms and water forms you listed on page 72. (Students' answers should reflect information appropriate to the expedition they selected.)

D. Describe some of the living things that you see on your journey. See the Almanac information on western trails for help with this. Take notes on the chart below. (Make sure students' descriptions reflect information on page 125 of the Almanac.)

People	Animals	Vegetation

E. Use the information you have gathered to create a painting or a drawing that shows the landscape and scenery you saw. Write a brief description to go with your painting or drawing.

(Students' paintings or drawings should appropriately reflect the expedition they selected.)

A. Use the map to complete the following items about the North and the South.

1. What were the original Confederate states?

South Carolina, Georgia, Florida, Alabama, Mississippi, Louisiana, Texas

2. Which border states joined the Confederacy?

Virginia, North Carolina, Tennessee, Arkansas

3. What were the original Union states?

California, Connecticut, Illinois, Indiana, Iowa, Kansas, Maine, Massachusetts, Michigan, Minnesota,

New Hampshire, New Jersey, New York, Ohio, Oregon, Pennsylvania, Rhode Island, Vermont,

Wisconsin

4. Which border states sided with the Union?

Delaware, Maryland, West Virginia, Kentucky, Missouri

5. How many states fought for the Union?

25

6. How many states fought for the Confederacy?

11

7. What city became the capital of the Union?

Washington, D.C.

8. What city was the capital of the Confederacy?

Richmond, VA

B. Think about the issue of slavery and the Civil War. Answer the following questions.

1. Did the North separate from the South or the South separate from the North?

The South separated, or seceded, from the North and formed the Confederacy.

2. Why were the states divided into the North and the South?

The South believed in slavery, but the North did not, so they fought the Civil War, which divided the country.

3. Was a larger area of the land occupied by people who were for or against slavery?

against slavery

77

Lesson 13

ACTIVITY Explore and express your opinion about the Civil War.

Civil War Resources

The BIG Geographic Question What situations did the northern and southern states face at the time of the Civil War?

From the article you learned that at the time of the Civil War, the North had resources that gave it many advantages over the South. The map skills lesson showed you how the country was divided into the Union and the Confederacy over the issue of slavery. Now write a newspaper article about this important time in America's history.

A. List six major kinds of resources that both the North and the South needed in order to fight the Civil War. (Answers may include the following.)

1. defendable land

2. industries

3. transportation

4. agriculture

5. military personnel

6. weapons and ammunition

B. Write a few sentences about one of the resources you listed above, comparing its availability and use in the North and the South and how it affected their readiness to fight a war.

(Students' answers should accurately describe and compare the availability and use of resources in

the North and South and how they influenced each side's readiness for war.)

C. Continue comparing the resources of the North and South by completing the following.

1. Name three industries that the North had.

(Possible answers include: shipbuilding, textiles, toolmaking, railroading, farming,

shipping, coal and iron mining.)

2. Why were railroads an important advantage for the North during the war?

(Students' answers should reflect awareness that the railroads opened more trading and industrial

opportunities and helped the war effort by moving troops and supplies more rapidly.)

3. List two major crops that were produced in each part of the country.

North	South
corn	cotton, tobacco
wheat	rice, sugar

4. Write a sentence describing one strength each side enjoyed at the start of the war. (Possible answers include the following.)

North: The North had many industries, including railroads, factories, and mining that could help

supply and transport goods needed for war.

South: The South held a strong commitment to its way of life which depended on slave labor,

and for which they would passionately fight.

D. Write an opinion editorial—a letter to a newspaper editor expressing your opinion about a topic—explaining who you think will win the war and why. Complete the following steps to help you organize your editorial. (Students' opinion editorials should include the characteristics of persuasive writing and who, what, when, where, and why details.)

1. Choose the viewpoint of either a northerner or a southerner at the beginning of the Civil War.

2. Include a headline that will capture your readers' attention.

3. Include in your first paragraph the important who, what, when, where, and why information.

4. Support your opinion with strong facts, reasons, and examples.

Lesson 14
MAP SKILLS
Using Maps to Compare
Urban Population Growth

Two maps can show the changes in population in a place over a period of time. Each map's legend, or key, can help explain the symbols and the colors used on the map. Notice that when maps are used to make comparisons, the legend is usually the same for both.

A. Study the maps' legends and answer the following questions.

1. What do the green dots represent? _population of 5,000 to 100,000 people in a city_

2. What do the red dots represent? _population of 100,000 people and over in a city_

B. Use information from the maps to answer the following questions.

1. By 1870, how many cities had a population over 100,000? _15_

2. By 1900, how many cities had a population over 100,000? _33_

3. By 1870, how many cities with populations over 100,000 were located on the Great Lakes? _2_

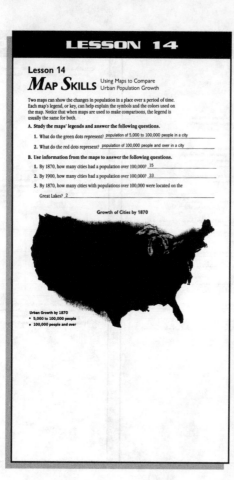

Growth of Cities by 1870

Urban Growth by 1870
- 5,000 to 100,000 people
- 100,000 people and over

4. By 1900, how many cities with populations over 100,000 were located on the Great Lakes?

7 (Detroit is not located directly on Lake Erie or Lake Huron)

5. By 1870, how many cities with populations over 100,000 were located on the Mississippi River?

_2_

6. By 1900, how many cities with populations over 100,000 were located on the Mississippi River?

_4_

C. Use the maps and the above information to help you draw some conclusions about how the nation's cities grew between 1870 and 1900.

1. By 1870, where were most of the cities with populations over 100,000 located?

in the Northeast

2. By 1900, which area of the country had experienced the greatest growth in cities with populations over 100,000?

the Midwest, especially around the Great Lakes and in the Ohio River Valley

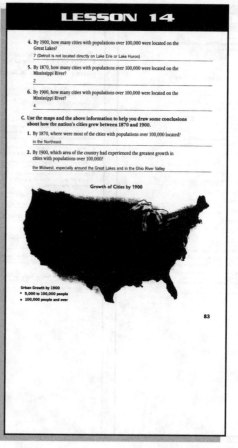

Growth of Cities by 1900

Urban Growth by 1900
- 5,000 to 100,000 people
- 100,000 people and over

83

Lesson 14
ACTIVITY
Find out about the effects of rapid growth on a city.

The Growth of a City

The BIG
Geographic Question

How did urban migration affect the city of Chicago?

You learned from the article that a great number of people moved from rural to urban areas in the United States after the Civil War. In the map skills lesson you looked at the growth in urban populations over time. Now find out how this enormous growth affected the city of Chicago.

A. Evaluate the rapid population growth of Chicago between 1860 and 1900. Use the article pages 80 and 81 to help answer the following.

1. What was Chicago's population by 1860? _109,260_

2. What was Chicago's population by 1900? _1,698,575_

3. Complete the following sentence about the population increase.

In _40_ years, Chicago's population increased by _1,589,315_ people.

B. Use information in the Almanac to complete the following chart about the many ethnic groups that migrated to Chicago between 1860 and 1900.

1. List several of the ethnic groups that migrated to Chicago.

Czechoslovakians	Russian Jews
Germans	Slovaks
Irish	Swedes
Italians	
Norwegians	
Poles	

2. What were some of the reasons immigrants came to Chicago?

They came to escape wars, persecution, overpopulation, and starvation in their homeland and to find a better life in America.

3. Why did African Americans leave the South and go to Chicago?

They came to find better jobs and wages.

4. Describe the area and housing in which immigrants and African Americans moving from the South lived in Chicago.

They lived in overcrowded tenement houses.

C. Describe how the land in Chicago was used during this time of sudden growth. Use the article, map, and Almanac to find out about transportation, parks, and modern conveniences in Chicago at the time.

When the immigrants flooded into Chicago, housing had to be built for them to live in. Chicago used a particular section of land in the city to build tenement houses and apartment buildings several stories high. Several people were crowded into each apartment and lived in poor and unsafe conditions.

D. Charles Dickens wrote in his book _A Tale of Two Cities_, "It was the best of times, it was the worst of times . . ." Using the above information you have gathered about Chicago, describe how this quote applies to Chicago's urban growth.

(Students' answers should reflect an understanding of how the growth of a city's population can boost the economy, with an increase in jobs and profits, but at the same time situations such as overcrowding and poor living conditions can lead to crime and unsafe streets.)

85

LESSON 15

A. List the major cities along the fall line.

Macon, GA — Raleigh, NC

Columbia, SC — Richmond, VA

Baltimore, MD — New York City, NY

Philadelphia, PA

B. Look at the map and write down the approximate elevations of the following cities.

1. Raleigh, North Carolina _0–500 feet_

2. Pittsburgh, Pennsylvania _1,000–2000 feet_

3. Baltimore, Maryland _0–500 feet_

4. Columbia, South Carolina _0–500 feet_

C. Look at the fall line on the map. Answer the following questions.

1. Explain why Raleigh, Baltimore, and Columbia are fall line cities.

Due to their low elevation and position on the Atlantic Coastal plain, these cities developed

along the fall line.

2. Explain why Pittsburgh is not a fall line city.

Pittsburgh is located in a hills region of western Pennsylvania on the western side of the Appalachian

Mountains. It has a high elevation. Fall line cities have lower elevations. They are located on the

Atlantic Coastal Plains.

D. In the twentieth century, industry has developed in all regions of the United States. Explain why industry is no longer dependent on physical features such as the eastern fall line.

Technology has advanced. We no longer have to depend solely on waterpower as an energy source.

LESSON 15

Lesson 15

ACTIVITY Find out how the Industrial Revolution was affected by the geography of the United States.

Geography and the Industrial Revolution

The BIG Geographic Question — What role did the eastern fall line and other geographical features play in America's Industrial Revolution?

From the article you learned how fall lines develop. In the map skills lesson you looked at fall lines and other geographical features on a relief map. Now learn more about what the eastern fall line had to do with the Industrial Revolution.

A. Answer the questions below to figure out how geography influenced the Industrial Revolution in the United States.

1. On the chart below list the cities that developed along the eastern United States fall line. Also, list the rivers associated with the eastern fall line.

Cities	Rivers
New York City	Hudson River
Philadelphia	Susquehanna River
Baltimore	Potomac River
Richmond	James River
Raleigh	Roanoke River
Columbia	Savannah River
Macon	Chattahoochee River

2. Why do you think New York is considered a fall line city?

New York City is at the mouth of the Hudson River. The river's source is the Appalachian Mountains.

LESSON 15

3. If you were building a factory in the 1800s, would you build it along the fall line or in a rural area? Explain your answer.

A fall line city would provide the volume of people needed to work in the factory and a

water transportation route needed to bring in raw materials and export products.

Also, the rivers would be an energy source for the factory.

4. What natural resources (rivers, minerals, vegetation) would you look for in an area where you were building your factory? Explain why each resource is useful.

(Possible answers include: river or seaport for transportation; water to power machines;

wood for construction.)

5. You have read about the effect of the eastern fall line on the industry of the region. Why did factories grow up along the fall line during the Industrial Revolution?

The waterfalls along the fall line provided a natural source of power to run the factory machines.

B. Make a plan for a 2- or 3-dimensional relief map that illustrates the eastern United States fall line. (Make sure students' maps include elevation and relief, mountains, and rivers that characterize the fall line.)

1. Write your plan below.

2. Decide what you will use to construct a raised relief map. You may use such materials as clay, aluminum foil, papier mâché, or construction paper. You may decide to use colors, shading, or contouring to show relief.

3. Construct your 2- or 3-dimensional relief map.

LESSON 16

A. Look at the map and key and answer the following questions.

1. What three mining resources does the map show?

coal, iron ore, limestone

2. What two ways might you be able to ship these mining resources?

by water and by railroad

3. If you were planning where to build a steel mill in Pittsburgh between 1890 and 1920, what would you need to consider?

where the necessary resources are located and how they could be shipped to your mill

B. Look at the location of the resources needed to produce steel and complete the following.

1. Describe the location of the iron ore deposits. Iron ore is located near Lake Superior in

Wisconsin and Minnesota.

2. Describe the location of the working coal fields. Coal fields are located near Lake Erie

in western Pennsylvania, eastern Ohio, and northern West Virginia.

3. Describe the location of limestone. Limestone is located south of Lake Erie, and

north of Lake Michigan or west of Lake Huron in northern Michigan.

4. What resource(s) would you need to transport and to what location?

The coal, iron ore, and limestone will need to be transported to the working coal fields.

5. How would you transport the resource(s)? They could be shipped over Lakes Superior, Michigan,

Huron, and Erie by boat or by railroad along the Great Lakes to the working fields.

C. Evaluate the information you have gathered above and decide where you would have built a steel mill in the early 1900s.

1. Design and draw a steel mill logo, or symbol, on the map to show the location of your mill.

2. Write an explanation of your decision below.

(Possible locations for mills include the Pittsburgh, Chicago, Detroit, Cleveland, and Buffalo areas.

Students' explanations should include the need for the mills to be near iron ore, coal, and limestone

as well as major shipping routes.)

LESSON 16

Lesson 16

ACTIVITY Find out about the movement of the steel industry over time.

Movement and Change

The BIG Geographic Question — Where was the United States steel industry originally located and why did it move over time?

From the article you learned how the steel industry began, grew, and declined. The map skills lesson showed you the location of the raw materials and the available shipping routes for Pittsburgh's iron and steel industry. Now find out what factors caused the location of the steel industry to change over time.

A. Use the article and Almanac to find out about various locations of the steel industry. Create a time line to show your information by completing the following.

1. From the 1860s to 1910

The steel industry was in Pittsburgh.

2. From 1910 to the 1920s

The steel industry shifted to the Lake Erie shores of Pennsylvania, Ohio, and Buffalo, New York.

3. From the 1920s to the end of World War II in 1945

The industry moved to the southern end of Lake Michigan, to Gary, Indiana, and Chicago, Illinois.

4. From the end of World War II to the 1970s

Many plants in the Northeast and Great Lakes areas closed down. The steel industry shifted to the

east and west coasts of the United States. Also, Japan and Germany started producing steel.

5. From the 1970s to the present

Many former steel-producing cities in the Northeast and Great Lakes areas have become service-

industry cities. A small amount of steel is still produced in the Northeast and Great Lakes areas.

LESSON 16

B. Use the article and Almanac to find information about the causes of change and movement in the steel industry in terms of the characteristics listed on the chart below. Complete the chart with the following questions in mind.

1. How did new technology increase the production of steel?

2. How did the way resources were used contribute to the movement of the steel industry from one location to another?

3. How did steel production in other countries affect the demand for U.S. steel?

Technology	Resources	Demand
• It was discovered that large amounts of steel could be made more quickly by blowing a blast of air through molten iron—the Bessemer process. • Factories' layouts were redesigned to be more efficient, and all components of steelmaking were centralized in a single facility.	• As soon as an ore deposit was depleted, new locations were sought. • Water provided a source of power and a means of transporting the steel.	• American iron and steel products are not competitive in world markets; therefore, the demand for American steelmaking has declined greatly.

C. Use the information you collected to create a visual display called "How the Steel Industry Has Changed over Time." Show the information from your chart in a time line, diagram, or poster of your own design. Use the space below to make a sketch of your visual display.

(Check students' visuals to see whether they effectively show how the steel industry changed over time.)

1. Using the article, the Almanac, and other reference materials, identify two advances in each field of technology and where they occurred. (Students' answers might include the following.)

Field of Technology	Technology Advancement	Location
Agriculture	cotton gin	New Haven, Connecticut
	barbed wire	De Kalb, Illinois
Transportation	steamboats	New York City, New York
	transcontinental railroad	Promontory, Utah
Communications	telegraph	Washington D.C. to Baltimore, MD
	telephone	Boston, Massachusetts
Air and Space	airplane	Kitty Hawk, North Carolina
	space shuttle	Cape Canaveral, Florida
Computers	CD-ROM	Redmond, Washington
	the Internet	Washington, D.C.

2. Create a symbol for each of the above fields of technology. Your symbols can consist of a shape, a picture, or one or two letters.
(Make sure students create symbols that are representative for each field.)
3. Draw your symbols on the map legend. Be sure to label the technology field that each symbol represents.
(Make sure students indicate which symbols represents which field on their map legend.)
4. For each invention you have listed, find its location on the map. Put a dot on the map where the city or town is located and label the dot with the city or town's name.
(Check to make sure students have accurately identified the locations of the inventions they included.)
5. At the correct location, draw the symbol for each field of technology. This symbol should be big enough to see clearly but small enough so that your map does not look too crowded. You may choose to draw your symbols in the open space bordering the map and draw a line from the symbols to the city locations on the map.
(Make sure the locations and symbols correspond according to students' choices.)
6. Analyze your map of inventions and discuss with classmates why you think the inventions were made in those locations.
(Help students discuss why they think the inventions occurred in these locations.)

Lesson 17
ACTIVITY Create a time line showing advances in technology.
A Technology Time Line

The **BIG** Geographic Question   How did technology advance in the United States?

From the article you learned how technology helped expand the borders of the United States. The map skills lesson helped you identify the locations of different inventions. Now make a technology time line to show when these inventions occurred. (Students may list any of the inventions mentioned in the article or Almanac.)

A. Make a list of important inventions that occurred between the 1700s and the present.

airplane   space shuttle

barbed wire   steamboats

cotton gin   telephone

Internet   transcontinental railroad

B. For the map skills lesson, you researched important inventions that helped the United States expand. You can use the same reference materials to find out the year when these inventions were created. Fill in the dates on the chart below. An example has been done for you. (Students' charts may include information on any of the inventions from the article, map skills lesson, and Almanac.)

Invention	Year Invented	Location	Significance
cotton gin	1793	New Haven, Connecticut	speeded up the removal of seeds from cotton so more could be produced and processed quicker

---

C. Now use the information you wrote on the chart to complete a technology time line. (Time lines should reflect information included in students' charts.)

1700   1793 1800   1900   2000   2100   2200

cotton gin

D. Evaluate the time line you completed and think about what new inventions will be created by the year 2100. (Paragraphs should name futuristic inventions and explain how they will affect the United States.)
1. Make a list of the inventions you think will be created in the future.

_____   _____

_____   _____

2. Use the space below to write a short paragraph about your predictions of inventions in the twenty-first century. Build the paragraph around the information that you collected on your charts and time line.

_____

_____

_____

3. Now add these invention predictions to your time line above.

4. Explain how these new technology advances will help the United States explore new frontiers.

_____

_____

_____

_____

_____

A. Look at the map of twentieth-century immigrants and the world map in the Almanac to complete the following.
1. List some of the countries in Europe from which immigrants came to the United States.

Russia, Germany, Great Britain, Norway, Poland, Italy

2. List some of the countries in Asia from which immigrants came to the United States.

Vietnam, Taiwan, China, Japan, Philippines

3. List some of the countries in the Americas from which immigrants came to the United States.

Mexico, Canada, the Caribbean, South American countries, Cuba, Panama, Puerto Rico

B. Look at the two graphs and answer the following questions.
1. According to the bar graph on the right, what was the total number of immigrants that came to the United States from the following continents between 1901 and 1990?

   a. Europe ___19.6 million___   b. Asia ___5.5 million___   c. the Americas ___11.5 million___

2. According to the line graph on the left, what was the total number of immigrants that arrived in the United States from the following continents between 1911 and 1920?

   a. Europe ___4.5 million___   b. Asia ___none___   c. the Americas ___1.5 million___

C. The line graph shows how things change over a certain period of time. It also lets us compare the number of immigrants from Europe, Asia, and the Americas. Use the line graph to fill in the blanks below.
1. Between 1901 and 1910, ___European___ immigrants were the most numerous group.
2. The ten-year period between ___1971___ and ___1980___ was the first time that there were more Asian immigrants than European immigrants.
3. Between 1981 and 1990, immigrants from ___the Americas___ were the most numerous group.
4. In the ten-year period between ___1931___ and ___1940___, immigration from Europe, Asia, and the Americas reached its lowest point.

Lesson 18
ACTIVITY Find out about the people who have immigrated to the United States.

Life in the Land of Opportunity

The **BIG** Geographic Question   Is the United States the land of opportunity for immigrants?

From the article you learned about some of the push and pull factors that brought many immigrants to the United States. The map skills lesson used a map and graphs to show the numbers of immigrants who came to the United States in the early twentieth century. Now find information about immigrants who have come to the United States in the 1990s, and conduct a mock interview to show whether immigrants still see the United States as the land of opportunity.

A. Complete the following about immigrants who came to the United States in the late nineteenth and early twentieth centuries.
1. List some of the factors that pushed immigrants from other continents and countries to the United States.

   a. civil war   d. poverty

   b. farmland shortage   e. religious persecution

   c. population pressure   f. starvation/famine

2. What were some of the opportunities they might have expected to find in the United States? (Answers may include religious freedom, large amounts of farm land available, plentiful jobs, and an abundant food supply.)

_____

_____

_____

_____

---

3. List some of the challenges that these immigrants faced once they arrived in the United States.

   a. different customs

   b. different language

   c. unemployment

   d. finding housing

   e. overcoming prejudice

B. Answer the following questions about people who are immigrating to the United States today.
1. Based on what you know about life today, do you think the United States is "a land of opportunity" for new immigrants coming here? Explain your answer.

(Students' answers should be supported by a description of current characteristics in the United States.)

_____

2. Who are some of the groups of people who are immigrating to the United States today?

(These groups include: Chinese, Israelis, Mexicans, and Bosnians. For a more complete list, see the listings in the Almanac.)

_____

_____

C. Interview an immigrant and compare his or her ideas about immigration with yours. Following are some questions you might ask.

1. Why did you leave your homeland?
2. How did you get to the United States? How long was your journey?
3. What opportunities did you expect to find in the United States? Did you find them?
4. Do you plan to stay or return to your home country?
5. Do you think the United States is the "land of opportunity"?

McGRAW-HILL LEARNING MATERIALS
Offers a selection of workbooks to meet all your needs.

Look for all of these fine educational workbooks
in the McGraw-Hill Learning Materials SPECTRUM Series.
All workbooks meet school curriculum guidelines and correspond to
The McGraw-Hill Companies classroom textbooks.

SPECTRUM GEOGRAPHY – NEW FOR 1998!

Full-color, three-part lessons strengthen geography knowledge and map reading skills. Focusing on five geographic themes including location, place, human/environmental interaction, movement and regions. Over 150 pages. Glossary of geographical terms and answer key included.

TITLE	ISBN	PRICE
Grade 3, Communities	1-57768-153-3	$7.95
Grade 4, Regions	1-57768-154-1	$7.95
Grade 5, USA	1-57768-155-X	$7.95
Grade 6, World	1-57768-156-8	$7.95

SPECTRUM MATH

Features easy-to-follow instructions that give students a clear path to success. This series has comprehensive coverage of the basic skills, helping children to master math fundamentals. Over 150 pages. Answer key included.

TITLE	ISBN	PRICE
Grade 1	1-57768-111-8	$6.95
Grade 2	1-57768-112-6	$6.95
Grade 3	1-57768-113-4	$6.95
Grade 4	1-57768-114-2	$6.95
Grade 5	1-57768-115-0	$6.95
Grade 6	1-57768-116-9	$6.95
Grade 7	1-57768-117-7	$6.95
Grade 8	1-57768-118-5	$6.95

SPECTRUM PHONICS

Provides everything children need to build multiple skills in language. Focusing on phonics, structural analysis, and dictionary skills, this series also offers creative ideas for using phonics and word study skills in other language arts. Over 200 pages. Answer key included.

TITLE	ISBN	PRICE
Grade K	1-57768-120-7	$6.95
Grade 1	1-57768-121-5	$6.95
Grade 2	1-57768-122-3	$6.95
Grade 3	1-57768-123-1	$6.95
Grade 4	1-57768-124-X	$6.95
Grade 5	1-57768-125-8	$6.95
Grade 6	1-57768-126-6	$6.95

SPECTRUM READING

This full-color series creates an enjoyable reading environment, even for below-average readers. Each book contains captivating content, colorful characters, and compelling illustrations, so children are eager to find out what happens next. Over 150 pages. Answer key included.

TITLE	ISBN	PRICE
Grade K	1-57768-130-4	$6.95
Grade 1	1-57768-131-2	$6.95
Grade 2	1-57768-132-0	$6.95
Grade 3	1-57768-133-9	$6.95
Grade 4	1-57768-134-7	$6.95
Grade 5	1-57768-135-5	$6.95
Grade 6	1-57768-136-3	$6.95

SPECTRUM SPELLING – NEW FOR 1998!

This series links spelling to reading and writing and increases skills in words and meanings, consonant and vowel spellings and proofreading practice. Over 200 pages in full color. Speller dictionary and answer key included.

TITLE	ISBN	PRICE
Grade 1	1-57768-161-4	$7.95
Grade 2	1-57768-162-2	$7.95
Grade 3	1-57768-163-0	$7.95
Grade 4	1-57768-164-9	$7.95
Grade 5	1-57768-165-7	$7.95
Grade 6	1-57768-166-5	$7.95

SPECTRUM WRITING

Lessons focus on creative and expository writing using clearly stated objectives and pre-writing exercises. Eight essential reading skills are applied. Activities include main idea, sequence, comparison, detail, fact and opinion, cause and effect, and making a point. Over 130 pages. Answer key included.

TITLE	ISBN	PRICE
Grade 1	1-57768-141-X	$6.95
Grade 2	1-57768-142-8	$6.95
Grade 3	1-57768-143-6	$6.95
Grade 4	1-57768-144-4	$6.95
Grade 5	1-57768-145-2	$6.95
Grade 6	1-57768-146-0	$6.95
Grade 7	1-57768-147-9	$6.95
Grade 8	1-57768-148-7	$6.95

SPECTRUM TEST PREP from the Nation's #1 Testing Company

Prepares children to do their best on current editions of the five major standardized tests. Activities reinforce test-taking skills through examples, tips, practice and timed exercises. Subjects include reading, math and language. 150 pages. Answer key included.

TITLE	ISBN	PRICE
Grade 3	1-57768-103-7	$8.95
Grade 4	1-57768-104-5	$8.95
Grade 5	1-57768-105-3	$8.95
Grade 6	1-57768-106-1	$8.95
Grade 7	1-57768-107-X	$8.95
Grade 8	1-57768-108-8	$8.95

Look for these other fine educational series available from McGRAW-HILL LEARNING MATERIALS.

BASIC SKILLS CURRICULUM

A complete basic skills curriculum, a school year's worth of practice! This series reinforces necessary skills in the following categories: reading comprehension, vocabulary, grammar, writing, math applications, problem solving, test taking and more. Over 700 pages. Answer key included.

TITLE	ISBN	PRICE
Grade 3 – new for 1998!	1-57768-093-6	$19.95
Grade 4 – new for 1998!	1-57768-094-4	$19.95
Grade 5 – new for 1998!	1-57768-095-2	$19.95
Grade 6 – new for 1998!	1-57768-096-0	$19.95
Grade 7	1-57768-097-9	$19.95
Grade 8	1-57768-098-7	$19.95

BUILDING SKILLS MATH

Six basic skills practice books give children the reinforcement they need to master math concepts. Each single-skill lesson consists of a worked example as well as self-directing and self-correcting exercises. 48pages. Answer key included.

TITLE	ISBN	PRICE
Grade 3	1-57768-053-7	$2.49
Grade 4	1-57768-054-5	$2.49
Grade 5	1-57768-055-3	$2.49
Grade 6	1-57768-056-1	$2.49
Grade 7	1-57768-057-X	$2.49
Grade 8	1-57768-058-8	$2.49

BUILDING SKILLS READING

Children master eight crucial reading comprehension skills by working with true stories and exciting adventure tales. 48pages. Answer key included.

TITLE	ISBN	PRICE
Grade 3	1-57768-063-4	$2.49
Grade 4	1-57768-064-2	$2.49
Grade 5	1-57768-065-0	$2.49
Grade 6	1-57768-066-9	$2.49
Grade 7	1-57768-067-7	$2.49
Grade 8	1-57768-068-5	$2.49

BUILDING SKILLS PROBLEM SOLVING

These self-directed practice books help students master the most important step in math – how to think a problem through. Each workbook contains 20 lessons that teach specific problem solving skills including understanding the question, identifying extra information, and multi-step problems. 48pages. Answer key included.

TITLE	ISBN	PRICE
Grade 3	1-57768-073-1	$2.49
Grade 4	1-57768-074-X	$2.49
Grade 5	1-57768-075-8	$2.49
Grade 6	1-57768-076-6	$2.49
Grade 7	1-57768-077-4	$2.49
Grade 8	1-57768-078-2	$2.49

THE McGRAW-HILL
JUNIOR ACADEMIC™ WORKBOOK SERIES

An exciting new partnership between the world's #1 educational publisher and the world's premiere entertainment company brings the respective strengths and reputation of each great media company to the educational publishing arena. McGraw-Hill and Warner Bros. have partnered to provide high-quality educational materials in a fun and entertaining way.

For more than 110 years, school children have been exposed to McGraw-Hill educational products. This new educational workbook series addresses the educational needs of young children, ages three through eight, stimulating their love of learning in an entertaining way that features Warner Bros.' beloved Looney Tunes™ and Animaniacs™ cartoon characters.

The McGraw-Hill Junior Academic™ Workbook Series features twenty books – four books for five age groups including toddler, preschool, kindergarten, first grade and second grade. Each book has up to 80 pages of full-color lessons such as: colors, numbers, shapes and the alphabet for toddlers; and math, reading, phonics, thinking skills, and vocabulary for preschoolers through grade two.

This fun and educational workbook series will be available in bookstores, mass market retail outlets, teacher supply stores and children's specialty stores in summer 1998. Look for them at a store near you, and look for some serious fun!

TODDLER SERIES
32-page workbooks featuring the Baby Looney Tunes™

	ISBN	PRICE
My Colors Go 'Round	1-57768-208-4	$2.25
My 1, 2, 3's	1-57768-218-1	$2.25
My A, B, C's	1-57768-228-9	$2.25
My Ups & Downs	1-57768-238-6	$2.25

PRESCHOOL SERIES
80-page workbooks featuring the Looney Tunes™

	ISBN	PRICE
Math	1-57768-209-2	$2.99
Reading	1-57768-219-X	$2.99
Vowel Sounds	1-57768-229-7	$2.99
Sound Patterns	1-57768-239-4	$2.99

KINDERGARTEN SERIES
80-page workbooks featuring the Looney Tunes™

	ISBN	PRICE
Math	1-57768-200-9	$2.99
Reading	1-57768-210-6	$2.99
Phonics	1-57768-220-3	$2.99
Thinking Skills	1-57768-230-0	$2.99

GRADE 1 SERIES

80-page workbooks featuring the Animaniacs™

	ISBN	PRICE
Math	1-57768-201-7	$2.99
Reading	1-57768-211-4	$2.99
Phonics	1-57768-221-1	$2.99
Word Builders	1-57768-231-9	$2.99

GRADE 2 SERIES

80-page workbooks featuring the Animaniacs™

	ISBN	PRICE
Math	1-57768-202-5	$2.99
Reading	1-57768-212-2	$2.99
Phonics	1-57768-222-X	$2.99
Word Builders	1-57768-232-7	$2.99

SOFTWARE TITLES AVAILABLE FROM McGRAW-HILL HOME INTERACTIVE

The skills taught in school are now available at home! These titles are now available in retail stores and teacher supply stores everywhere.
All titles meet school guidelines and are based on
The McGraw-Hill Companies classroom software titles.

MATH GRADES 1 & 2

These math programs are a great way to teach and reinforce skills used in everyday situations. Fun, friendly characters need help with their math skills. Everyone's friend, Nubby the stubby pencil, will help kids master the math in the Numbers Quiz show. Foggy McHammer, a carpenter, needs some help building his playhouse so that all the boards will fit together! Julio Bambino's kitchen antics will surely burn his pastries if you don't help him set the clock timer correctly! We can't forget Turbo Tomato, a fruit with a passion for adventure who needs help calculating his daredevil stunts.

Math Grades 1 & 2 use a tested, proven approach to reinforcing your child's math skills while keeping them intrigued with Nubby and his collection of crazy friends.

TITLE	ISBN	PRICE
Grade 1: Nubby's Quiz Show	1-57768-011-1	$19.95
Grade 2: Foggy McHammer's Treehouse	1-57768-012-X	$19.95

MISSION MASTERS™ MATH AND LANGUAGE ARTS

The Mission Masters™ -- Pauline, Rakeem, Mia, and T.J. – need your help. The Mission Masters™ are a team of young agents working for the Intelliforce Agency, a high level cooperative whose goal is to maintain order on our rather unruly planet. From within the agency's top secret Command Control Center, the agency's central computer, M5, has detected a threat… and guess what – you're the agent assigned to the mission!

MISSION MASTERS™ MATH GRADES 3, 4 & 5

This series of exciting activities encourages young mathematicians to challenge themselves and their math skills to overcome the perils of villains and other planetary threats. Skills reinforced include: analyzing and solving real world problems, estimation, measurements, geometry, whole numbers, fractions, graphs, and patterns.

TITLE	ISBN	PRICE
Grade 3: Mission Masters™ Defeat Dirty D!	1-57768-013-8	$29.95
Grade 4: Mission Masters™ Alien Encounter	1-57768-014-6	$29.95
Grade 5: Mission Masters™ Meet Mudflat Moe	1-57768-015-4	$29.95

MISSION MASTERS™ LANGUAGE ARTS GRADES 3, 4 & 5 – COMING IN 1998!

This new series invites children to apply their language skills to defeat unscrupulous characters and to overcome other earthly dangers. Skills reinforced include language mechanics and usage, punctuation, spelling, vocabulary, reading comprehension and creative writing.

TITLE	ISBN	PRICE
Grade 3: Mission Masters™ Feeding Frenzy	1-57768-023-5	$29.95
Grade 4: Mission Masters™ Network Nightmare	1-57768-024-3	$29.95
Grade 5: Mission Masters™ Mummy Mysteries	1-57768-025-1	$29.95

FAHRENHEITS' FABULOUS FORTUNE

Aunt and Uncle Fahrenheit have passed on and left behind an enormous fortune. They always believed that only the wise should be wealthy, and luckily for you, you're the smartest kid in the family. Now, you must prove your intelligence in order to be the rightful heir. Using the principles of physical science, master each of the challenges that they left behind in the abandoned mansion and you will earn digits to the security code that seals your treasure.

This fabulous physical science program introduces kids to the basics as they build skills in everything from data collection and analysis to focused subjects such as electricity and energy. Multi-step problem-solving activities encourage creativity and critical thinking while children enthusiastically accept the challenges in order to solve the mysteries of the mansion. Based on the #1 Physical Science Textbook from McGraw-Hill!

TITLE	ISBN	PRICE
Fahrenheit's Fabulous Fortune	1-57768-009-X	$29.95
Physical Science, Grades 8 & Up		

All titles for Windows 3.1™, Windows '95™, and Macintosh™.

Visit us on the Internet at

www.mhhi.com